T0256149

Andean Meltdown

Andean Meltdown

A CLIMATE ETHNOGRAPHY OF WATER, POWER, AND CULTURE IN PERU

Karsten Paerregaard

UNIVERSITY OF CALIFORNIA PRESS

University of California Press
Oakland, California

© 2023 by Karsten Paerregaard

Cataloging-in-Publication Data is on file at the Library of Congress.

ISBN 978-0-520-39391-2 (cloth)
ISBN 978-0-520-39392-9 (pbk.)
ISBN 978-0-520-39393-6 (ebook)

32 31 30 29 28 27 26 25 24 23
10 9 8 7 6 5 4 3 2 1

To my brother

CONTENTS

ILLUSTRATIONS

MAPS

CHART

FIGURES

ACKNOWLEDGMENTS

I have spent almost four decades collecting the data that I review in this book and spent several years writing about it. To name all the people who have contributed to my work and helped me conduct field research is therefore not possible. However, I want to thank everyone—dead or alive—who has supported—wittingly or unwittingly—or taken part in—directly or indirectly—my endeavors to write this book.

The research would have been impossible without the generous grants from the Danish Research Council, the Swedish Research Council, and Sweden's Riksbankens Jubileumsfond. I started to write the book during a stay as visiting researcher at the Center for Latin American Studies, University of Florida, where director Philip Williams and many old friends and colleagues provided me with invaluable support and encouragement. Another piece was written at the Section for Global Development, Department of Food and Resource Economics, University of Copenhagen, where head of section Christian Lund and his colleagues received me with open arms and provided me office facilities. The manuscript was completed at the Swedish Collegium for Advanced Study, which hosted me for an academic year as a research fellow and has been a critical source of motivation and inspiration. A special thanks to principal Christina Garsten and the collegium's staff. I'm also indebted to my colleagues at the School of Global Studies, Gothenburg University, who always have been supportive of my work.

Collaborating with students and colleagues in the field has been an invaluable experience while gathering field data for the book. I'm particularly indebted to Astrid B. Stensrud, Astrid O. Andersen, Susan B. Ullberg, and Malene K. Brandshaug for the wonderful years we've spent together in Arequipa and the Colca Valley conducting fieldwork, individually as well as collectively. Dream teams do not exist only in basketball! A special thanks to Mattias B. Rasmussen, with whom I've shared many great moments—at home as well as abroad—and who gave me invaluable comments on the manuscript.

In Peru numerous people have contributed to the book. Teófilo Altamirano has always been there to discuss my ideas and experiences. Many thanks to him and his family! I'm also indebted to colleagues at the Department of Social Sciences, Pontificia Universidad La Católica del Perú, for their support. In Huancayo I'm grateful to Carlos Condor, Universidad Nacional del Centro del Perú, who introduced me to the wonders of Huaytapallana. My thanks also go to Pedro Marticorena, who welcomed me at his museum. In Cusco I want to thank Pablo Concha Sequeiros for accompanying me to Quyllurit'i and sharing time with me in the city of Cusco. I also appreciate the incredible support that people in Tapay and Cabanaconde have provided me over the years. A special thanks to my compadres Rufina Riveros and Mauricio Llasaca in Cosñirhua for our lifelong friendship and to my old friend Godofredo Riveros and his wife in Tapay for letting me participate in the offering to Mount Seprigina. In Cabanaconde I'm indebted to Saida Valdivia for assisting me with the surveys I conducted in the community and to Juan Valdivia for inviting me to join his excursions to Mount Hualca Hualca.

As always, I'm indebted to my wife, Ana María Torres, her family in Lima and elsewhere, and our two daughters for their support and their trust in my work.

Finally, I want to thank my editors at the University of California Press, Stacy Eisenstark, Kate Marshall, and Naja Pulliam Collins, for the support they have provided me. On a similar note, I'm grateful for the critical as well as constructive comments and suggestions the two external reviewers gave me.

Parts of the book's case studies have previously appeared in journals or edited volumes. Smaller pieces of the ethnographic descriptions in chapters 2, 3, and 4 have been published in *Religions* 4 (2): 290–305; *Environmental Communication* 14 (1): 112–125; *Water Alternatives* 12 (3): 488–502; *kritisk etnografi: Swedish Journal of Anthropology* 4 (12): 83–98; *Climate & Development* 10 (4): 360–368; *Mountain Research and Development* 33 (3): 9–12; *WIRE's Water* 5 (2): 1–11; and *Anthropology and Nature*, edited by K. Hastrup (Routledge, 2014, pp. 196–210). Chapter 4 is a revised and extended version of a chapter in *Understanding Climate Change through Religious Lifeworlds*, edited by D. Haberman (University of Indiana Press, 2021, pp. 261–283), while chapters 4 and 5 comprise ethnographic data previously published in *HAU: Journal of Ethnographic Theory* 10 (3): 844–859.

Introduction

"Great *apu*, we have come here to greet you and show you respect," said the shaman while raising his arms and looking toward Mount Huaytapallana, which houses the *apu*, the nonhuman being embodied in the mountain. The crowd of about a thousand pilgrims who had gathered around the shaman imitated his gesture of saluting Huaytapallana. The solemn atmosphere lasted several minutes, after which people started to embrace and wish each other a happy new year. Then the forceful sound of a piece of ice breaking off the glacier interrupted the cheerful mood, reminding people not only of Huaytapallana's spiritual power but also of the threat of global warming and the glacier melt it causes. Unnerved by the sound of the calving glacier, a woman asked, "What will happen to Huaytapallana when the ice is gone?" while a man next to me exclaimed, "Huaytapallana has heard our prayers." When I asked the pilgrims about their view of the mountain's melting glacier, one person said, "The *apu* is dying," while another claimed "it is a *pachakuti*," a Quechua term for the turning of the world upside down.[1]

This account is an extract from my field diary in 2014 from when I took part in the annual celebration of the Andean New Year at the foot of Huaytapallana's glacier in the central highlands of Peru. The experience made an unforgettable impression on me and induced me to write this book. Who is the *apu*? Why is it dying? Why does its future death produce a *pachakuti*? And more broadly, how do global climate change and the retreat of Huaytapallana's and the rest of Peru's tropical glaciers affect the culture and ritual customs of the country's Andean population?[2] By addressing these and other related research questions, the book contributes to the growing body of anthropological research on climate change. Using case studies from four field sites in the Peruvian highlands, it offers an ethnographic account of how Andean people interpret and make sense of climate change and how this becomes a part of their daily lives, prompting them to reinvent social practices and reshuffle their worldviews. My argument is that rather than viewing climate change as an isolated or external phenomenon, Andean people experience it as one of many forms of change occurring in their lives. Moreover, even though Andean people are among the first to suffer from climate change, many do not view this as anthropogenic, and those who do think it is caused by their own activities, not by the agency of people in other parts of the world (Jurt et al. 2015; Paerregaard 2019b).[3] It is therefore my contention that climate change and its impact on the society and culture of Andean people must be investigated within a broader context of environmental tensions, economic development, social conflict, and cultural change at the local, regional, and national levels.

As one of the most densely inhabited mountain regions in the world, the Andes is extremely sensitive to global warming, which causes glacier melt, flooding, and water scarcity and leads to environmental degragation, social conflicts, and out-migration. Using Peru as an example, my research suggests that climate change represents both a dilemma and a possibility for countries in the Global South that suffer from water scarcity.[4] Rising temperatures, irregular precipitation, flooding, and water shortages pose fundamental environmental problems for Peru.[5] But climate change also challenges the country's hierarchical structure by urging the state to include its marginal communities in climate adaptation projects. At the same time, climate change opens the door for new forms of political engagement by inciting people to review the human-environment relationship and the ideas of control and dominance that inform the state's politics. In other words, climate change, glacier melt, and water scarcity transform Peruvian society in multiple ways. On the one hand, they cause social conflicts, fuel internal migration, and jeopardize economic growth; on the other, they pave the way for new forms of social inclusion and citizenship and offer vulnerable populations a new perspective on Peru's environmental politics and the development model underpinning its prosperity (Paerregaard, Stensrud, and Andersen 2016). The book's proposal is that this tension between anthropogenic change and environmental crisis versus social inclusion and political mobilization constitutes an inherent paradox not only in Peru's current efforts to reduce poverty and raise the living standards of people in marginal areas while mitigating the impact of climate change but also in climate change politics in other parts of the world.

Most of the existing literature on climate change within the social sciences focuses on the immediate consequences of global warming (Yearly 2009). Employing such concepts as resilience, adaptation, and vulnerability, this first generation of climate change scholarship examines how exposed communities adjust to rapid environmental change (Adger 2006; Adger et al. 2003; Folke 2006; Oliver-Smith 2016) and how external agencies assist them in implementing emergency plans and inventing coping strategies to overcome natural hazards caused by climate change (Oliver-Smith and Shen 2009). Anthropologists in particular offer a comprehensive understanding of social reliance and vulnerability as embedded in everyday human agency because they apply a broad, holistic view of human and natural systems and study within the shifting contexts of environmental and social change (Crate 2011). Anthropologists also work directly with people who are affected by this change and are exposed to the new climate realities (Hastrup 2013), which enables the researchers to identify how global warming affects locations where it is being felt most urgently and examine how people experience and adjust to climate change locally (Crate and Nuttall 2009, 2016).

While the strength of anthropology is to capture the subtle ways people deal with climate change in their local life worlds, it also recognizes that these are never homogeneous, isolated, or static, but form part of the larger world. Modern anthropologists are therefore adept at examining long-term environmental change, contextualizing this in national and global perspectives, and embracing several scales in the study of climatic and social change (Barnes et al. 2013; Barnes and Dove 2015; Hastrup 2016). Writing on behalf of the emerging community of anthropologists studying climate change, Kirsten Hastrup contends: "We must learn to theorize across ethnographic fields and offer our theories to the wider community of scholars and scientists for inspection and inclusion in the general field of climate change research" (2016, 36). A growing number of anthropologists have taken up this challenge of studying climate change as a multiscale phenomenon (Bauer and Bhan 2018; Bear and Singer 2014; Greschke and Tischler 2015) and exploring how it is experienced, anticipated, and perceived by people and becomes part of their everyday lives in a variety of places (Paerregaard 2016; Stensrud 2016a, 2016b; Strauss and Orlove 2003). Recent anthropological studies have also investigated how climate change challenges ideas of nature and culture and instigates people to rethink their relationship to the environment and interpret meteorological phenomena (Ingold 2007; Paerregaard 2014b), how people predict the local implications of climate change and model nature's course (Hastrup and Skrydstrup 2012), and how global warming produces new patterns of human mobility (Hastrup and Olwig 2012).

One of the questions addressed by the anthropological literature is the conflictive nature of not only climate change but also climate change research, which in many places is an issue of dispute and contestation (Hulme 2009). Even though climate change is global and impacts all corners of the world, it does not manifest itself as a distinct and independent phenomenon (Calder 2015). As Heike Greschke points out: "Despite being regarded as a serious problem for all humans in present and future times, climate change is not directly perceptible. Knowledge about the causes and effects of global warming has to be mediated and can only become socially relevant at particular sites if it connects to life experiences and culture-specific patterns of interpreting the environment" (2015, 123). Consequently, people's perceptions of climate change are shaped by their own experiences and cultural ideas and are often at variance with the science-based and Western-generated global discourse on climate change, which separates the epistemic from the normative and detaches global facts from local value, destabilizing knowledge at the same time that it seeks to stabilize (Adger et al. 2012; Crate 2011; Jasanoff 2010; Mathur 2015; Paerregaard 2018a).

Elaborating on this observation and its implications for the dissemination of climate knowledge, Werner Krauss and Hans von Storch write: "The communication between climate science and the general public is severely disturbed" (2012, 214). The cause of this discord, Krauss and von Storch assert, is that "global climate models and their regional counterparts neither reflect nor match the climate reality people inhabit" (2012, 214). To conceptualize this discrepancy and understand why climate change has become a political battleground and a key narrative within which all environmental politics is now framed, Krauss and von Storch describe climate change research as a postnormal science.[6] Such a research approach differs from traditional scientific practice, demanding what Krauss and von Storch call "an extended knowledge basis"—that is, the inclusion of the social and cultural disciplines and the voices of the people they work with in climate research (2012, 226). In other words, to engage with society and make its results available to nonprofessionals, climate change science must collaborate with the social sciences and the humanities and make use of their qualitative-oriented methods and frameworks to examine climate change as not only a physical but also a social and cultural phenomenon (Moulton et al. 2021). Ethnographic studies bring home this point by showing how global discourses on climate change intertwine with indigenous cosmology, local morality, and national politics, and how this merging of different knowledge systems generates unexpected and controversial ideas about the human-nature relationship and the causes of global warming (Brügger, Tobias, and Monge-Rodríguez 2021; Burman 2017; Crona et al. 2013; Green and Raygorodetskty 2010; Paerregaard 2013a).

But as a postnormal science, climate change does not only question the nature of its facts, values, stakes, and urgency. It also raises fundamental questions about humanity and its role in planet Earth's future prospects. The scientific community now overwhelmingly attributes global climate change to human activities, prompting many to employ terms such as *anthropogenic* and the *Anthropocene* to underscore humans' double role as both a main contributor to and a steward of the planet's climatic and environmental problems (Steffen, Crutzen, and McNeill 2007). In the words of Hastrup: "Humans are everywhere, not only as destroyers of nature but also as providers of collective solutions" (2016, 36). The acknowledgment of humans' pivotal role for the planet's evolution speaks to the heart of anthropology and induces anthropologists to engage in climate research by inquiring into the multiple ways people explain climatic change, particularly how they account for their own contributions to its cause and effect (Greschke 2015; Jurt et al. 2015; Paerregaard 2020a; Schnegg, O'Brian, and Sievert 2021). The questions such an anthropogenic research focus asks include: Who are the "we" in the Anthropocene? And how do we distribute blame and guilt in the discussion of what has caused anthropogenic climate change? As Sayre points out: "The politics of the anthropogenic must give way to a politics that identifies which people have caused which changes, with what consequences to whom, and demands a justice that is indistinguishably social and environmental at the same time" (2012, 67).

Some scholars, however, find that conventional political thinking fails to tackle the underlying problems of anthropogenic climate change, which, according to historian Dipesh Chakrabarty, produces a crisis in the distribution of natural reproductive life on the planet. Chakrabarty argues that "our political and justice-related thinking remains very human-focused" and asserts that "we still do not know how to think conceptually—politically or in accordance with the theories of justice—about justice towards nonhuman forms of life, not to speak of the inanimate world" (2017, 32). Therefore, Chakrabarty writes, there is an urgent need for a politics of the

Anthropocene that reaches beyond conventional understanding of "the political" as a mere human affair and addresses anthropogenic climate change and, in particular, what he calls the Great Extinction—that is, human-driven extinction of other species on a massive scale and other irreversible human footprints on Earth's system.

Chakrabarty's critique of the politics of the anthropogenic echoes recent anthropological works on posthumanism that both inspect the human/nonhuman relation as a multispecies engagement (Aisher and Damodaran 2016; Kirksey and Helmreich 2010), a human-nature collaboration (Choy et al. 2009; Tsing 2015), a multinatural lifeworld (Latour 2011), and an ecology of mutually constitutive materials (Ingold 2012) and call for a revision of the universality and the notion of a unified cosmos implied in conventional politics, whether practiced by the Right or the Left (Latour 2013). This critique also resonates with the notion of environmental cosmopolitanism, as suggested by Ben Campbell (2008), and the idea of a cosmopolitics, as proposed by Isabelle Stengers (2010, 2011), that regards the cosmos as an unknown and open space of divergent worlds and explores the possibility of articulating them with each other to become a common world (Blaser 2016, 546–547).[7] Such a cosmopolitics recognizes that the world is more than one socio-natural formation: in Marisol de la Cadena's words, a "kaleidoscopic simultaneity of similarity and difference" (2015, 22). And while a cosmopolitics aims to interconnect its multiple forms of existence, it does not treat them as commensurable (2015, 22). As de la Cadena writes, "A new pluriversal political configuration—perhaps a cosmopolitics, in Stengers' terms—would connect different worlds with its socionatural formations—all with the possibility of becoming legitimate adversaries not only within nation-states but also across the world" (2010, 361).[8]

Unlike scholars who scrutinize cosmopolitics as a project that starts as a theoretical claim and aims to disrupt established ways of thinking politics, I approach it as an empirical phenomenon that emerges from humans' experience and interpretation of their own anthropogenic agency and that can both coproduce and alter conventional political practices (Paerregaard 2019c). As demonstrated by de la Cadena (2015) in her study of mountain deities and other earth beings in the Andes, indigenous ritual practices and cultural imaginaries speak to and defy the established rules of political engagement at one and the same time. But climate change not only reveals the blind spots of conventional politics; it also complicates humans' contributions to cosmopolitics. As my ethnographic case studies show, the encounter with glacier retreat and chronic water shortage challenges Andean people's own understanding of their relationship with the earth beings they believe control the water flow and questions their notion of what it implies to be human.

My bottom-up approach to cosmopolitics has implications for the way I conceptualize the Andean pluriverse and, as I discuss in the section on data collection, the way I position myself in the field. For years Andean anthropologists posited the society-environment nexus as a divide between two separate worlds, one exclusively human and the other, labeled "nature," comprising all other forms of existence. While their works are full of ethnographic accounts of how the line between the two realms are blurred in ritual practices, symbolic representations, and mythical configurations, such crossings of the human-nature divide are described as activities and ideas that unfold and exist in people's cultural world rather than in the real world (Abercrombie 1998; Allen 1988; Bastien 1978; Bolin 1998; Gose 1994; Isbell 1978). More recently, a growing number of anthropologists have taken issue with this approach. Questioning the opposition between humans and the environment that underpins this approach and putting its notion of a "pure" natural world under arrest, they argue that reality or nature emerge from rather than precede human practice (Blaser 2013; de la Cadena 2015; Descolá 2013; Ingold 2012; Latour 2013). Instead of drawing on the "ethic/emic" framework that anthropologists conventionally have used to distinguish their own perspectives from those of their interlocutors, these scholars employ the term *ontology* to describe people's embeddedness in the environment they inhabit, claiming that anthropologists should take their interlocutors' ideas of nature at face value and recognize them as being as valid as their own (Descolá 1996; Ingold 2007; Latour 2011). Some scholars even propose an indigenous cosmopolitics that departs from a reality that is constituted by indigenous people's own concept of the world and that may give rise to a *political ontology*: a hegemonic struggle of defining and creating the world (Blaser 2016; Burman 2017; de la Cadena 2010).

While supporting the effort to break up the society-nature divide and welcoming the invitation to acknowledge the epistemological value of indigenous (as well as other) people's *worldings*—that is, their way of inhabiting and perceiving the world (de la Cadena and Blaser 2018)—my approach is pragmatic. More specifically, to study the cultural impact of climate change in the Andes, I borrow from both the conventional understanding of mountains and other nonhuman agents inhabiting the environment as symbolic representations in Andean ritual practice and cosmology and the new reading of these forces as possessing agency and being real. Employing the two theoretical positions as complementary rather than exclusive approaches, I argue that the mountains attain different meanings in different settings. In some contexts, they are best understood as figures of existential importance in people's lives and livelihoods that demand recognition as material beings acting as agents on a par with humans. In other contexts, they should be approached as symbolic configurations that may be critical for people's cultural practices and ideas but that nevertheless are issues of dispute and objects of negotiation and contestation and therefore cannot be dealt with unequivocally as self-contained, autonomous agents. I develop this proposition further in the four case studies, which illustrate how the social status that people attribute to mountains varies both within the same setting and across regions, sometimes appearing as metaphorical representations, other times as material beings.

My overall argument is that by making humans mindful of their own position in Earth's system and the impact their activities have on it and of their role as a planetary agent in relation to other life-forms, firsthand experiences of rapid climatic change upset people's perceptions of nature and their ideas of what are figurative characters and what are real agents in the environment. But climate change not only undermines local worldviews and epistemologies; it questions science's authority and calls for a new cosmopolitics (Paerregaard 2020b). Or as Candis Callison puts it: "Climate change cuts to the core of who and what human concerns are and how they are mediated and moralized. It enables questions beyond what the realm of science offers: What is our relation to each other, locally and globally? What is our relationship to the earth?" (2014, 23). The species identity emerging from such a climate consciousness is neither exclusive nor stable and may coexist with and even coproduce existing identities based on national, ethnic, or cultural belonging. Rather than replacing traditional kinds of intrahuman politics and existing forms of cosmopolitics, the awareness of living in an anthropogenic world adds a new dimension to environmental, social, and indigenous politics, in some situations transforming it and in others merely transfiguring it.

There is not one but many answers to the planet's environmental problems, and even though these become evident to many when they witness hard-core climatic facts, people's stakes and options in an anthropogenic world differ just as their possibilities of responding to its challenges vary

(Beck 2010; Emmett and Lekan 2016). By drawing the attention to humans' responsibility for glacier retreat and the world's water crisis, cosmopolitics undercuts fixed ideas of what is human and nonhuman. However, in doing so it also affirms the terms of ordinary politics by bringing to the fore the social and economic inequalities climate change glosses over. An ethnographic study of the vernacular experience of rapid glacier retreat and water shortage in mountain regions offers a look into the moral and cultural landscape that frames the global discourse on climate change and highlights the predicaments that impel people struggling to adapt to its consequences to simultaneously contest and abide by the established rules of political engagement.

WATER METABOLISM

Just as I examine how mountains and water sometimes are objects of interpretations and configurations and at other times emerge as beings in the real world, I scrutinize how people both construct water as a cultural image and engage with it as a substance endowed with life and agency. Borrowing from the growing body of anthropological literature on water's social nature (Attala 2019; Beresford 2020; Orlove and Caton 2010; Paerregaard 2018b; Strang 2005, 2015), I explore on the one hand how water fashions Andean people's worldview and interaction with nature and, on the other hand, how Andean culture and ritual practice shape their adaptations to climate change and the water crisis it causes. To unpack the human-water nexus in the Andes and the ideas that drive Andean offerings, I investigate these as a replicate of the metabolic process by which material objects change chemical composition and physical form and, as a result, produce energy and life.[9] A key concept in this proposal is water metabolism, which is derived from the notions of social metabolism and the hydrosocial cycle and which I employ with two classics in anthropology in mind, one old and one more recent: Karl Marx and his notions of metabolism, human alienation, and second nature, and Roy Rappaport and his proposal to study rituals as a regular part of the human-nature relationship.

Around 150 years ago Marx wrote that the ideologists of bourgeois society had created a false opposition between nature and humans and criticized the notion of humans as a species alienated from nature, free to exploit its physical environment (Marx 1992). As a politically engaged intellectual, Marx pointedly drew attention to social inequality and environmental pollution, which he claimed were caused by capitalist production. Just as the workers lived under the yoke of economic profit, so had nature become the slave of humans, he contended. But Marx also claimed that the relations of exploitation under capitalism are based on an immanent contradiction between the technological and scientific development and the social order, and that once the former has undermined the latter, both humans and nature will be able to obtain freedom. In his praxis theory he fleshed out the inconsistencies of capitalist production, reminding us that nature and society are inextricably linked together and that human beings, like society, are an integral, yet particular and radically distinct, part of nature (Swyngedouw 2006, 108). Employing the notion of metabolism, he scrutinized nature as the material in which human labor realizes itself. Marx's understanding of metabolism was closely linked to the term's German meaning, "change of matter" (*stoffwechsel*), which implies a continuous process of transforming and reassembling of material elements (Swyngedouw 2006, 108). Thus in Marx's view, through the mediation of labor, society emerges from nature, resulting in the production of a "second nature,"—that is, the reassemblage of human-material objects resulting from human labor. Marx's definition of the nature-society nexus as a metabolic relation led him to assert that "the workers can create nothing without *nature*, without the *sensuous external world*" (1992, 325) and to conclude that "nature is man's *inorganic body*, that is to say nature in so far as it is not the human body" (1992, 328). As laborers, then, humans can only bring the wrongdoings of capitalism and their own self-alienation to a stop by engaging with the second nature as their external body.

Marx's writings have received renewed interest in a time of global climate change that transforms the planet into an anthropogenic world and that lends humans a feeling of double alienation: first their separation from the "first nature" they have ruthlessly exploited and then their separation from their self-produced, unrecognizable "second nature" that now is turning against them (Escobar 1999). Marx's concepts of metabolism and alienation have a particular bearing on anthropology because they help understand the human-nature nexus.[10] Yet anthropology also challenges Marx's theory by showing that humans' engagement with nature covers a complex relationship that historicizes the landscape and extracts specific places out of undifferentiated space (Lowell 1998, 6). Moreover, in many societies nature constitutes not only the "sensuous external world" and "inorganic body" of their members in their daily struggle to satisfy the physical needs but also the central point of reference of their worldview, providing them with a sense of place and locality and a feeling of belonging. To anthropology, therefore, Marx's metabolism refers to a biological and sociopolitical as well as a cosmological and metaphorical relation of exchange; similarly, his concept of alienation can be read as at once a critique of capitalism and modern consumer society and a lens to explore humans' relationship to nature and their position in the physical environment more broadly.

In fact, Marx's idea of metabolism speaks to a long tradition in anthropology of studying the human-nature nexus and environment-society-culture dynamics. In the history of environmental anthropology, one work stands out as particularly relevant to modern anthropologists: Rappaport's (1968) ethnographic account of the Kaiko ritual and its role in regulating both intra-human and extrahuman relations among the Maring people of New Guinea. Influenced by systems theory, cybernetics, and nutritional science, Rappaport showed how Kaiko and the slaughtering of pigs it involved marked both the time of warfare with neighboring groups and a shift in horticultural production, which is essential for the management of the ecosystem. His argument that Kaiko enabled Guinea people to restore the human-pig ratio and create environmental sustainability was groundbreaking insofar as it demonstrated that culture may play an active role in managing humans' interactions with their physical surroundings. Rappaport's approach to ecological metabolism and sustainable animal farming, however, was premised on a notion of the closed local community and a neglect of the power relations that underpin the cultural-ecological equilibrium obtained through the Kaiko ritual. Such a lack of scrutiny of the structural conditions that frame and the external forces that shape the ethnographic setting is incompatible with contemporary anthropology in general, but it is particularly critical for the research on climate change that aims to uncover the global, national, and regional dynamics driving change in small-scale communities, an issue I return to in chapter 1.

But even though Rappaport's study was constrained by his view of culture as a self-producing system and lack of a global perspective, it deserves renewed attention because it exemplifies how anthropology can contribute to ongoing debates on human/nonhuman relations at a moment of global climate change. Aletta Biersack writes: "While Rappaport explicitly undertook to demonstrate human adaptations to a nature that stood outside the human realm, it is clear in retrospect that what he actually offered in *Pigs for the Ancestors* was an ethnography of nature, as it were: as study of the intersection of culture and nature, rooted as this intersection is in human activity, conceptualization, values and social relations" (2006, 7). Paradoxically, at a later stage in his writings Rappaport's functionalistic understanding of culture and reductive notion of human agency led him to a fatalistic view of cultural adaptation, which he argued could result in "maladaptation" and a breakdown of the ecosystem. Like Marx, who coined the phrase "metabolic rift" to draw attention to humans' overexploitation of the soil and the commercial use of nonhuman manure to meet the growing demand for food products, Rappaport thus anticipated the possibility of an ecological crisis even though he had no way of predicting the scope of climatic and environmental problems humans would face fifty years after he was writing.[11]

The proposal to study humans' exploitation of nature as a metabolic disruption and to examine culture as a metabolic regulator is particularly relevant in water studies. Influenced by political ecology and theoretical trends within geography, environmental social studies, and cognate disciplines, water scholars have recently questioned the hydrologic cycle as a neutral scientific concept, arguing that it is a social construct that has emerged in a specific historical context and in response to economic and political needs.[12] This scholarship draws our attention to the relationship between water and society and the way socioeconomic and political forces modify the circulation of water, creating what these scholars call a "hydrosocial cycle." Asserting that we must examine water and society as mutually constituting rather than pregiven phenomena, Jamie Linton and Jessica Budds define the hydrosocial cycle as "a socio-natural process by which water and society make and remake each other over space and time" (2014, 170). In a similar vein, Rudgerd Boelens explains that "hydrosocial cycles are simultaneously natural and social constructs—as chains of human and nonhuman elements constructed by the human mind and by human interventions" (2014, 245).

To understand how hydrosocial cycles generate a specific instance of water forming what Linton and Budds describe as "an assemblage of historical, hydrological, political and technological circumstances" (2014, 2017), another group of scholars suggest that we view the circulation of water as a metabolic process produced by not only physical changes and chemical reactions, as modern science claims, but also social and political relations that seek to manipulate the natural water flow and transform water into "H_2O"—that is, a resource for human use (Madrid-López and Giampietro 2015). Unlike the term *metabolic water*, which refers to water created inside a living organism through its metabolism, *water metabolism* stands for the entire process through which water is made and remade, both as a natural substance (from liquid to ice and vapor and back again) and as an object of human control (from appropriation to transportation, allocation, consumption, and eventually, disposal). The concept originates in the same idea as the hydrosocial cycle: that society shapes and is shaped by water, both materially and discursively, and that water flows are embedded in all institutional and political processes that both coexist with them and affect them (Swyngedouw 2009). However, water metabolism adds analytical value to the notion of the hydrosocial cycle by highlighting the social practices humans engage in and the relations they create with their biophysical surroundings when appropriating water and transforming it into H_2O. Rather than understanding water as a mere resource flow, as a factor of production only disaggregated from its environmental, institutional, technological, and social context, as the notion of the hydrological cycle does, water metabolism views it as an integral element of a metabolic process organized around humans' appropriation and transformation of nature (Beltrán and Velásquez 2017).

If metabolism served as key metaphor for Marx's definition of labor and his critique of capitalism, and if rituals occupied a central role as the regulator of humans' interactions with the environment in Rappaport's theory, water metabolism can enable me to scrutinize Andean rituals as an act that triggers a "change of matter" of the offering items and that transforms these into a gift to the mountain deities that Andean people believe control the circulation of water (de la Cadena 2015; Paerregaard 1989; Salas Carreño 2016; Salomon 2018; Stensrud 2016a, 2021). Moreover, using Marx and Rappaport helps me understand how notions of power are displayed in Andean rituals and how their representations of authority change meaning and form as climate change, and the glacier melt and water scarcity it causes, undermine the quest that drives them: to smooth the hydrological cycle and generate more water. More bluntly, water metabolism is a window on Andean people's struggle to make sense of climate change and adapt to its consequences (Paerregaard 2016). Last, as discussed later, I use the concept to frame the book's ethnographic case studies, which explore four ways Andean people reshuffle their cosmology and adjust their ritual practices in response to the growing water scarcity. My point is that while water rituals in Andean communities are tied to the existential need for water and embedded in local imaginaries of submission and accountability to the powers that control the water flow, they are attributed symbolic rather than material importance in Andean pilgrimages, which mostly are attended by people from urban settings and increasingly are regulated by regional, national, and global institutions.

To explore anthropologically how climate change is experienced as a multiscale phenomenon and intersects with environmental, social, and political change, regionally as well as nationally and globally, Susan Crate suggests a framework for a climate ethnography, which she argues must entail "the development of a new multisited, critical collaborative ethnography that integrates perceptions, understandings, and responses by both modifying resilience/adaptation frames and further developing cultural models" (2011, 185). My study adopts Crate's proposal by using data from four separate but thematically connected field sites, of which each constitutes what Marcus calls a strategically situated ethnography. Such a climate ethnography is a foreshortened multisited project that attempts to understand something broadly about the world system and current globalization processes in ethnographic terms as much as it does its local subjects (Marcus 1998, 95). More specifically, a strategically situated ethnography identifies places that are of strategic relevance to the chosen topic of research and that allow the researcher to use local insights to shed light on issues of global importance such as climate change (Paerregaard 2008a). For an Andean climate ethnography, this means to document Andean people's efforts to adapt to climate change in a selection of field sites that in distinct but complementary ways reveal something important about the way ongoing processes of economic, political, and social change in Peru fashion their experience, interpretation, and response to melting glaciers, irregular precipitation, growing water scarcity, and other forms of environmental change. Equally important, the field sites, which are located in three geographically different regions of the Andes, bring to the fore critical variations in Andean water scarcity and water management, the two first being located in Peru's southwestern highlands, where water is chronically scarce and where most crops are irrigated; the third in its central highlands, where precipitation is less irregular and agriculture is to a large extent rainfed; and the fourth in its southeastern highlands, where water is more plentiful and where irrigation is less common.

Reviewing ethnographic data from four strategically situated research sites in the Peruvian highlands (see map 1), the book brings to the fore the paradoxes and predicaments Andean people face when they adjust their social organization, cultural worldview, and ritual practice to climate change. Two of the four sites are in Peru's southwestern highlands and offer an everyday perspective on how Andean people experience and respond to climate change at a local level. Tapay and Cabanaconde are neighboring communities that have

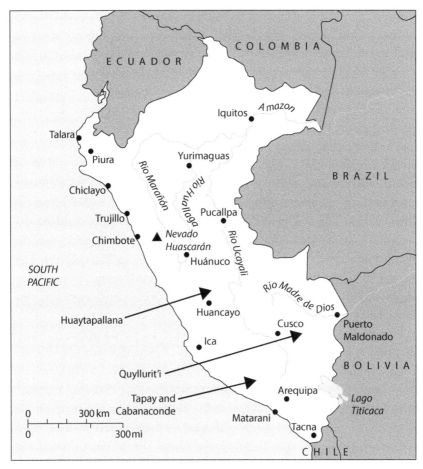

MAP 1. Map of Peru showing the four field sites. Map downloaded and redrawn from Mother Earth Travel, https://motherearthtravel.com/peru/map.htm.

chosen different strategies to alleviate their water shortages but nevertheless face the same challenge: to adapt to the water crisis by rearranging their water supply chain; rethinking their notions of water, power, and community; and seeking new political alliances. The two other sites are in Peru's central and southeastern highlands and describe how people adapt to climate change at a regional and national level. Mount Huaytapallana and Mount Qulque Punku (also referred to as Quyllurit'i) are settings of annual pilgrimages that differ in terms of historical development, social structure, and religious practice but nevertheless are caught in the same dilemma: to practice the Andean worldview and belief in mountains without harming the environment.

The analytical thread that ties the four cases together is my analytical focus on how climate change and the glacier melt and water shortage it causes challenge Andean ritual practices. In the first two cases the ice-water-culture nexus serves as an analytical lens to examine how climate change undercuts the cosmology and offerings of Peru's highland communities and prompts them to revisit their view of the state, private mining companies, and other external agents. The last two cases, on the other hand, use the same nexus to explore how climate change creates not only a concern for Peru's glaciers and environment but also a renewed interest in Andean cosmology and ritual practice in Peruvian society at large. The following overarching questions guide the four case studies. How does climate change challenge the communities' and the pilgrims' perception of water, snow, and ice, and the mountains they believe control these elements? How do the communities and the pilgrims adjust their worldview and the ritual practices associated with it to the idea of an anthropogenic world? How do external agents influence the communities' and the pilgrims' ability to adapt to climate change, and how do the latter respond to such interventions?

I develop my climate ethnography in three steps. First, I inspect the four field sites as acts of habitation and instances of social life that bind people into communities of shared ethical criteria and commitments and provide them with a feeling of social belonging (Lambeck 2011). Second, I explore them as settings for cultural events and collective activities that tie people to specific places in the landscape and mediate their contact with the non-human beings inhabiting it. Third, I examine how the changing physical and sociopolitical contexts of the four field sites transform these practices and life-forms and challenge the worldview they are anchored in. The aim of this approach is to "field climate change," as suggested by Carla Roncoli, Todd Crane, and Ben Orlove (2009), by describing the social relations and moral ideas that shape Andean people's experience and interpretation of climatic and environmental change and to reconstruct their interaction and communication with both the state and its institutions and the metaphysical powers in the Andean landscape. The following questions guided my research: Why do some communities continue to pay tribute to the mountains they believe control the water while other communities cease doing it? How does the changing ritual practice of the latter affect their worldview, and how do the former deal with the fact that the water source they are paying tribute to is drying up? Why do a growing number of people from urban settings suffering from water shortage participate in the regional pilgrimages to the region's highest mountains? How do the pilgrims' firsthand experiences of glacier retreat and environmental change during these pilgrimages challenge their religious ideas and notions of anthropogenic climate change? How do Andean people envision a world without glaciers and ice?

Employing water metabolism as an analytical concept to describe how Andean people adapt to climate change by revisiting their relationship with mountains and engagement with water, I frame the four case studies as distinct pathways to rework the metabolic assemblage of ice, water, humans, and power. But while my ethnographic scrutiny aims to illustrate the distinct ways this reassemblage is practiced in different regions and contexts, the pathways should not be read as models that fit neatly into the four cases. On the contrary, two or more pathways may converge in the same location, just as the same pathway can be pursued in different situations. Consequently, analogies between them can be drawn by myself as a researcher as well as by the people I work with in the field. To demonstrate how the perception of mountains and water alters regionally and contextually, I present offering rituals as a continuum of metabolic calibrations: the offering is perceived at one extreme as a material gift to a living being and at the other extreme as a metaphorical tribute to an imagined power. My point is that while the former predominates in the first case study and the latter prevails in the fourth case study, the two calibrations converge in the second and the third case studies, though in different amalgamations.

Labeling the first study "the existential metabolism," I suggest offerings in the community of Tapay are crafted as a reciprocal relationship between humans and mountains that commits the former to bringing the latter gifts in exchange for releasing the water, a resource of vital importance to the community. The case study exemplifies how an Andean community that relies on a nearby mountain as its principal water source manages this and how it continues to conduct offerings to appease the mountain, which the bulk of villagers view as a real being, even though a small minority of villagers fiercely deny its existence. At the same time, the case study discusses the dilemmas the community faces because a mining company that operates within its territories has offered to sponsor a new water infrastructure. While the mining company's proposal lends some water users an illusion of progress, it has stirred the concern of others about the community's environment and future water supply and spurred yet others to accuse the community leaders who are negotiating the deal with the company of lacking integrity and legitimacy. If Tapay's negotiation with the mining company bridges over internal discords on the effectiveness of ritual offerings and the prospects of the community's water supply, Cabanaconde's modernization of its water governance and abandonment of collective offering rituals are the consequence of its confrontation with the state forty years ago to access water from a state-built channel called Majes. I call this second case study "the contractual metabolism" because it examines how the state has offered the community a contract that promises it a stable and safe water supply in exchange for paying a water tax and a water tariff. To highlight the circumstances that have made Cabanaconde change from the "existential metabolism" that previously prevailed in the community to "contractual metabolism," the chapter describes the dramatic circumstances that first made the state grant Cabanaconde the right to water from the Majes channel and later induced the community to replace its own model of water management with a state model. Even so, the villagers have started to make offerings to the channel, suggesting that the two metabolic calibrations now coexist in the community.

The third case examines the processes of negotiation and contestation that motivate the pilgrims of Huaytapallana to review the meaning and purpose of their offering gifts and to reflect on the harm these do to the mountain. I call this case "the transactional metabolism," which refers to the notion of offerings as quid pro quo deals that allow the givers to specify the goods or favors they expect the receiver to deliver in exchange for their gifts. Even though the fungible nature of the offerings urges the pilgrims to view the mountain and the water as symbolic powers rather than as living beings, the mountain's vital role as water supplier for the nearby city induces some to ascribe the mountain and water agency and life. The chapter's focus is on the controversies that the offering practices and the harm these do to the environment trigger in a situation where people perceive climate change as a local rather than a global phenomenon and as an outcome of their own rather than other people's agency. Finally, I explore Quyllurit'i as an event that not only enables pilgrims to make sense of the glacier retreat and its repercussions for their activities and beliefs but also urges them to mobilize against intruding mining companies and adjust to the regulations the state imposes on the pilgrimage. I dub the pilgrimage "the metaphorical metabolism" because its ritual practices seek to meet spiritual rather than physical needs and because the tributes their followers make are figurative rather than material. Presented as the fourth and last case study, Quyllurit'i is emblematic of the symbolic representation of mountains and water flows. And while the young men who walk up to the glacier and engage in physical contact it undoubtedly experience both the mountain and the ice as living beings, the bulk of pilgrims, who watch the ritual from afar, regard the mountain and the glacier as symbolic powers rather than agents bestowed with real life. The chapter brings to the fore the double pressure Andean pilgrims come under when climatic change compels them to adapt their ritual traditions to a world without glaciers at the same time that external actors interfere in their activities, encroach on their territory, and profit from their traditions.

DOING FIELDWORK

While the book's case studies can be read as four individual ethnographic descriptions of how climate change shapes the lives of Andean people, they follow a single analytical script, which falls into two parts. The first part examines the impact climate change has on glaciers and water resources at the community level, where water is valued because of its material quality and viewed as a question of life or death, and the second looks at the regional level, where water is attributed symbolic meaning and where national and global actors increasingly are making their presence felt. But the case studies are also a reflection of my own travelling in the Andes and represent an account of how some of the extensive fieldwork I have conducted in Peru during the past four decades has developed over time. My journey has been a personal as well as anthropological voyage, from a local and confined community (Tapay), where I know just about everybody, to first a bigger and more complex location (Cabanaconde), where I feel familiar but only know a few, then an urban/regional environment (Huaytapallana), where I formerly had conducted fieldwork but now engage in an entirely new research issue, and finally to a national/global setting of which I have no previous experience (Quyllurit'i).

Situated opposite each other in the bottom of the Colca Valley in the country's southwestern highlands, the communities of Tapay and Cabanaconde have been my principal field sites throughout this period. In 1986 I conducted fieldwork for one year in Tapay, which lies on the northern side of the Colca River, scrutinizing such issues as bartering and ecological verticality, household and community organization, ritual practices and religious movements, and rural-urban migration.[13] I followed up on this research in 1989 by inquiring into Tapay's migrant colonies in the cities of Lima and

Arequipa and the economic and social relations that tie them to their community of origin, a study that took me almost two years. Over the following fifteen years I did fieldwork on a smaller scale several times in Tapay and visited the community on a regular basis to keep in contact with people.[14] Alarmed by the news about glacier retreat in Peru, I returned to Tapay in 2011 to document how the community was coping with its climate vulnerability and shrinking water supply, a topic I have pursued on annual visits over the past eight years. Located on the southern side of the Colca River, Cabanaconde, on the other hand, remained a secondary field site for several decades. As the principal point of entrance to Tapay, I have recurrently visited Cabanaconde, where I know many villagers. Yet it was only in 2007 that I started to focus my research on Cabanaconde, exploring two issues. The first is Cabanaconde's transnational migration network and its migrant colony in Washington, D.C., and the second is the community's water management and the villagers' climate perception and climate vulnerability.[15]

My field methodology in Tapay and Cabanaconde encompasses formal interviews with community leaders and ritual experts and informal conversations with villagers of both sexes and different generations. I have also interviewed directors and engineers from the regional and national water institutions in Arequipa on several occasions and participated in their meetings with community leaders from Tapay and Cabanaconde. In my 1986 study of Tapay I collected life stories of a selected group of villagers, conducted a village census, and conducted a household survey with questions regarding economic activities, irrigation practices, family relations, and community participation. In my restudy in 2011 I employed a new survey with one-fourth of the households in Tapay's two largest hamlets on climate perceptions. I complemented the survey with informal interviews with fifteen villagers in the same two hamlets on what they believe causes environmental and climate change. I also collected life stories of eight elderly villagers with a focus on climate change. With the assistance of a villager, Saida Valdivia, I conducted two surveys between 2011 and 2012 in Cabanaconde; the first yielded data on the community's socioeconomic situation and the second on the villagers' climate perception. From 2014 to 2016 I followed up on the surveys by interviewing community authorities and individual villagers on the challenges that climate change poses to Cabanaconde. My study of Cabanaconde also draws on Paul Gelles's community study from 1987 (Gelles 2000) and my personal conversations with the author about the community and its migrant colony in Washington.

My inquiry into Andean pilgrimage reviews ethnographic data from field-work between 2014 and 2017.[16] I have participated three times in the annual pilgrimage to Mount Huaytapallana on June 21, in 2014, 2016, and 2017, each time accompanied by Carlos Condor, a local anthropologist and old friend of mine. Before and after the events I interviewed the organizers and principal participants in the pilgrimage and conducted informal discussions with some of the pilgrims. The theme of these conversations was people's motives to participate in the pilgrimages, their perception of the relation-ship between humans and mountain deities, and their interpretation of climate change and its importance for their future lives. On my last trip to Huaytapallana, I was accompanied by a journalist and photographer from American University in Washington, D.C., Bill Gentile, who made a docu-mentary video about the pilgrimage.[17] In 2016 I participated in the pilgrim-age of Quyllurit'i, spending two cold nights at the pilgrim site of Sinak'ara together with Pablo Concha Sequeiros, also an anthropologist and an old friend of mine, who previously had made the pilgrimage and who assisted me in administering a questionnaire among the pilgrims who constitute Quyllurit'i's organizational core.[18] In 2017 I returned to the city of Cusco, where I participated in a demonstration the pilgrims had arranged to protest against the Ministry of Mining and its plan to grant a mining company con-cession to operate within the pilgrimage's territory. In the days that followed I interviewed the principal stakeholders in the pilgrimage about climate change and other issues related to its future prospects.

Doing long-term ethnographic fieldwork implies personal interactions and commitments with the communities you work with. I have already de-scribed how my relationships with people in the book's four field settings have developed over time. In the same period, anthropology's concept of objectivity and subjectivity and notions of ethical conduct have undergone a radical change. In the early 1980s the fieldworker was still regarded as a lone wolf whose individual attributes and individual conduct rarely were accounted for in the ethnographic account. Today, positionality and trans-parency are key words in the work of the anthropologist. To meet these imperatives, the book highlights my engagement with the people I worked with as well as my relationships with local researchers. Because I have made numerous trips to especially the communities examined in the two first case studies, and because the number of people I have met and known during the past forty years are almost uncountable, there is no way I can account for all my encounters and acquaintances. Nevertheless, the book provides portraits and conveys voices of some of my key interlocutors and Peruvian colleagues who have collaborated with me in the field.

Equally important for ethnographic writing is the recognition of people's ideas and worldviews even when these diverge from the received wisdom or the scholar's own notions. As already discussed, rather than taking a theoretical position a priori by applying a conceptual framework that explores mountains and water as either symbolic representations or ontological beings, I start from the ground listening to what people say and observing what they do and then adapt my analytical tools to their experiences of offering rituals and interpretations of the more-than-human world that inhabits the environment. What matters to critical ethnography is to provide an account of how people's actions, relations, and ideas are embedded in their own lifeworld rather than using the field data to pick an argument over theorical positions.

STRUCTURE OF THE BOOK

The book is informally organized in three parts. The first serves as a theoretical, methodological, and contextual framework for the study. In addition to this introduction, it includes chapter 1, which offers a brief review of how Andean religious imaginaries have intersected with power relations and social identities in the pre-Hispanic, colonial, and republican eras, and in particular how Andean people have adapted to shifting political regimes by reshuffling their worldview and incorporating images of their changing masters in their cosmology. I also discuss the impact global climate change is having on Peru, environmentally as well as politically, and the challenges it poses to Andes culture and identity. Finally, I introduce the book's case studies and discuss how I employ the notion of water metabolism to examine the offerings' ceremonial repertoire, cultural imaginary, and environmental implications and the role they play in the different strategies Andean people pursue to adapt to climate change.

The second part comprises two of the four ethnographic case studies examining the challenges Andean communities face as their water supplies shrink. Chapter 2 examines "the existential metabolism," which is the case study of the community of Tapay, while chapter 3 scrutinizes "the contractual metabolism," which describes the development of the community of Cabanaconde.

The third part shifts scale from the local to the regional and national levels and presents the two case studies that deal with Andean pilgrimages and their struggles to adapt ritual practices to climate change. Chapter 4 interrogates "the transactional metabolism," discussing the pilgrimage of Mount Huaytapallana, while chapter 5 investigates "the metaphorical metabolism," exploring how Quyllurit'i has become one of Peru's most popular pilgrimage sites.

In the conclusion I sum up the study's results, historically and socially comparing and contextualizing its ethnographic insights and reflecting on their theoretical and political implications. I identify the common thread of the four case studies and discuss how climatic and political change interlock in Peru. The conclusion also inquires into the social and cultural relations that make climate change and glacier melt an issue of conflict in Peru and instigate people to mobilize. Finally, it considers the bearings the book's findings have on our understanding of social and cultural change in Andean society and more broadly the humanities' conceptualization of the relations between climate, power, and culture and the importance of human activity and creativity for planet Earth's future evolution.

ONE

Water, Power, and Offerings

In chemical terms, ice, water, and vapor are the same substance. Nonetheless, their different physical appearances make them look like distinct elements. As Luci Attala points out, "water cannot successfully be thought of as one thing. It's composition changes constantly, as do its shape and density. Consequently, water should be more accurately represented as a transformational substance that shifts from one manner of being to the next" (2019, 19). As a liquid, water is subject to the law of gravity, which makes it flow downhill anywhere it exists. In effect, water is untamable unless it is exposed to heat, which transforms it into vapor and makes it disappear into the air. Ice, on the other hand, is water in a solid state that appears both fixed and stable; that is, until it melts and becomes water or vapor. Ice's stability, however, varies in different parts of the world depending on the latitude and the temperature. In the arctic, ice is permanent and covers both the land and the sea most of the year. By contrast, in the tropics ice is a rare phenomenon only found at high altitudes and out of human reach, which lends it an image of not only aesthetic beauty but also infinity and mystery.

The appearance of ice as a substance that is inaccessible and that belongs to another physical realm makes it a compelling motive for cultural creativity and religious adoration. The imaginary power of ice is particularly salient when it appears in the form of glaciers, which are huge, mobile ice formations that grow on the top and melt at the bottom at one and the same time and that constitute the source of some of the worlds' biggest rivers and a vital freshwater deposit in regions suffering from irregular precipitation and water scarcity (Allison 2015; Orlove, Wiegandt, and Luckman 2008). Not surprisingly, the capacity to reproduce themselves while yielding a constant flow of clean water makes glaciers the object of not only political craftsmanship and social empowerment but also cosmological ingenuity and ritual activity, which the Andes is evidence of (Bolin 2009; Paerregaard 2018b).[1] Hence the questions guiding my inquiry: What relations of power and domination are at play in this world of mountains, deities, humans, and water? How do humans' agency and their interaction with nonhuman forces influence the water flow in the Andean cosmology? How do Andean people imagine offerings as a means of smoothing the hydrologic cycle and water's metabolic process?

Before answering these questions, I offer a historical introduction to the symbolic representation of water, ice, and mountains in Andean politics and cosmology to help us understand the complex relationship between ritual and power in modern-day offering practices and the changes they currently are undergoing due to climate change and water scarcity.

INCAS, SPANIARDS, AND *HUAKAS*

Capturing, containing, and controlling water is a difficult task anywhere, but in the Andes it is particularly demanding because of the altitudes and deep slopes and because much of the water available comes from glaciers and ice- and snowcaps at high altitudes above 5,000 meters. At the same time, most Andean crops rely on irrigation, which makes water management a fundamental aspect of social organization and a key to power in the region (Boelens 2015; Gelles 2000; Guillet 1992; Mitchell and Guillet 1994; Rasmussen 2015; Stensrud 2021; Trawick 2003; Treacy 1994). As the first pan-Andean empire, the Inca state commanded the administrative capacity and mustered the resources and labor needed to construct large water infrastructure. The Incas built their hydraulic expertise on astronomic wisdom and climatic knowledge, which their programmers and engineers used to anticipate water's transformations and its future appearances and flows (Sherbondy 1994; Zuidema 1986). Knowing how to assess, tame, and manage water was essential not only to meet material needs but also to craft and command political power (B. Lynch 2019). Water therefore constituted a fundamental element in the Inca cosmology, which construed the circulation of water as a relation of reciprocity between humans and a nonhuman realm of mountains, ancestors, and other agents.[2] By picturing these as effigies of human agents, the Inca cosmology blurred the line between the sacred power, which the forces prevailing in the metaphysical world hold, and the social and political power, which the actors ruling in the human world command. And to legitimize their own sovereignty, the Incas represented themselves as human proxies of the divine force that created the world and therefore as masters of the water flowing from the mountains' ice and their springs.

The blend of adoration of ice and water, on the one hand, and forefather cult, on the other, which constituted the Inca cosmology was fueled by the idea that people could trace their ancestors to specific mountains and that these required offering gifts, sometimes in the form of human sacrifices, in return for releasing the water they controlled (Besom 2013; MacCormack 1991). Thomas Besom explains that the Incas founded their political power on ceremony, and to exercise authority over the subjects of their empire they "relied heavily on rites involving human sacrifice and mountain worship" (Besom 2013, 20). According to Besom, the Incas usurped and aggrandized the worship of peaks in the provinces they conquered as a means of tying the native populations to the Inca polity (2013, 54). Another method the Incas utilized to naturalize their authority involved "projecting the concept of social hierarchy onto the landscape" (2013, 54) based on the notion that geographical altitude symbolizes political and religious power. In this hierarchy, the most prominent mountain both in terms of its sacredness and its visual impact "would be associated with the lion's share of sacrificial materials" (2013, 55), while "at the bottom of the hierarchy would be hills that were venerated by the indigenous people and that received only locally produced materials" (2013, 55). Of critical importance to this configuration of the Andean landscape was the sacrifice of different classes of humans to specific mountains (2013, 39), which the Incas used to forge a hegemonic structure that visualized the social order of their empire and in particular the relations of dominance between the conquerors and the conquered. Andean geography was essential to this structure and the ideology of mountain offerings it relied on. Placed on ice-capped peaks or glaciers above 5,500 meters, the victims of Inca sacrifice remained intact as frozen bodies for centuries, reminding the local population of the conquerors' almighty power.

The Spanish invasion and defeat of the Inca Empire in 1532 and the following subjection of Andean people to colonial rule, which lasted almost three hundred years, stands as the quintessence of a disruption. It led to a demographic disaster that reduced the Andean population dramatically in the first fifty years and prompted the Spanish administrators to resettle the survivors in so-called *reducciones* modeled as replicates of Castilian towns (Cook 1981; S. E. Ramirez 1996). The Spaniards also made Andean people toil in the mines and submitted them to a rigid and demanding tax regime (Cook 2007; Stern 1993). Furthermore, they forced the native population to convert to Christianity and launched aggressive campaigns to eradicate the Incas' ancestor cult and their adoration of and offerings to mountains that included human sacrifices (MacCormack 1991). Yet even though the conquered adopted the conquerors' language, culture, and religion and only offered their rulers occasional resistance, the preconquest heritage continued to pervade the social organization, cultural habits, and belief system of Andean society and culture. The cultural and religious syncretism that emerged from this coexistence of Andean and Iberian heritage was furthered by the conquerors' use of the Incas' religious practice to validate their dominance (Sallnow 1987, 21–25). According to Peter Gose, the Spaniards made "a pact of reciprocity" with the Andean population that enticed the latter to pay tribute to their new masters in exchange for the latter's recognition of their communal land tenure (Gose 2008, 7). Moreover, though banning human sacrifice and the adoration of mummies and frozen bodies and forcing the indigenous population to convert to Christianity, the Spaniards placed Christian symbols on top of Andean holy sites, thus recognizing their continuous importance. The conquered, on the other hand, adapted to colonial rule by identifying the new conquerors with their ancestors and the mountains, thus incorporating the conquerors in the Inca cosmology (2008, 6–7). According to Gose, "Converting intrusive colonists into indigenous ancestors was an attempt to domesticate their power and make it serve indigenous interests" (2008, 6). The outcome of this syncretism was a colonial society that was Spanish and Christian by name and in form but that in practice and substance continued to be Andean (Abercrombie 1998, 258–282; Gareis 2019; MacCormack 1991, 406–433).

Drawing on the Incas' anthropomorphic representation of the sacred and their representation of as themselves as proxies of the superhuman powers, Andean people thus adapted to colonial rule by reinventing their forefathers in new religious garments, now as Spaniards. This reshuffling of the Andean pantheon established a genealogical linkage between the old and new rulers and created a sense of continuity with the pre-Spanish past. In the words of Gose, Andean people domesticated the Spanish conquerors by identifying the conquerors with their own ancestors and incorporating them in their cosmology. As a result, *huakas*—holy places in the landscape where the Inca and pre-Inca population kept their ancestors' bones and important relics such as ceramics and textiles—continued to serve as sites of ritual gatherings and cultural identification for Andean people during the colonial period. The religious syncretism emerging from the blend of European and American axioms and customs allowed the Andean people to keep their religion alive. It also enabled them to incorporate the Inca idea of reciprocity into the Catholic universe of symbols and icons and continue the practice of paying tribute to nonhuman agents such as mountains and forefathers. But to Andean people these offerings were not only a gift to the extra-human powers in return for the use of natural resources such as water, land, and minerals they controlled in accordance with their pre-Hispanic culture. They were also a tool to communicate with the Spanish rulers and a ritual compass to navigate in the social and political space of the colonial state. In other words, offerings became a medium by which the conquered population engaged in a symbolic relation of exchange with the conquerors to negotiate the terms of their submission.

ANDEAN REVIVAL

Andean people have continuously "updated" the syncretized postconquest cosmology by retailoring the human characters of its pantheon to make them match with their new rulers. When Peru gained independence in 1821, colonial rule was replaced by republican governance, which in ideology included all Peruvians but in practice excluded the indigenous population and created a divide between a dominating stratum of Hispanic descent and a dominated stratum of pre-Hispanic descent. Rather than abolishing the Spanish system of feudal privileges and its indexation of its subjects in castes and racial groups, Peru's new rulers "Indianized" the country's rural population by portraying them as cultural "others" and external political subjects (Gootenberg 1991; Kubler 1952; Thurner 1997). The republic's new class structure and division into ethnic groups emerged in the institutional vacuum Peru's independence created in the Andean highlands, which provincial authorities and hacienda owners used to establish a regional power regime known as *gamonalismo* (Mariátegui 1974; van den Berghe and Primov 1977). Key to the *gamonales'* power was their control of vital resources such as labor, land, and water and a hegemonic rule that represented the hacienda owners, labeled mestizos, as the descendants of Peru's Spanish conquerors, and the peasants and rural workers, dubbed *indios*, as the ignorant and uncivilized descendants of the Inca population.

To adjust to *gamonalismo*, Andean people reset their cosmological compass by replacing their pre-Hispanic ancestors with mountains configured as *apus*, which they viewed as a more-than-human power with the ability to control the flow of mountains' meltwater.[3] Even though Andean people continued to adore *huakas* and other nonhuman agents such as *pachamama* (Mother Earth) during the republican period (and even though they also paid tribute to mountains in the colonial era), the *apus* now became a central point of reference in their cosmology and the idea of human-nature reciprocity. And just as Andean people has previously recognized the conquerors as their rulers in the image of their pre-Hispanic forefathers, they adapted to the republic's new power constellation by identifying the hacienda owners and the regional mestizo class they were part of with *apus* and therefore as their new masters. The image of the *gamonal* and his associates (lawyers, regional officials, priests, etc.) as mestizo and its conflation with the adoration of the *apu* continued to dominate throughout the first half of the twentieth century and was pivotal to the perpetuation of the *misti*'s (Quechua: mestizo) opposite ethnic image: the *indio*.[4] Associated with the *apu*, the *misti* was feared for his power to command and punish but also respected for his capacity to protect and yield favors. And in the image of the *apu*, the *misti* demanded gifts in return for such services, which became an ideological cover to gain access to land, water, and other resources and to cloak the relations of exploitation and abuse that reigned in the highlands throughout most of the twentieth century.

The *gamonalismo* began to crumble in the mid-twentieth century, but in some places it lasted up to the 1969 land reform, which finally dismantled the haciendas and reformed Peru's land tenure system (Mayer 2009; Paerregaard 1987b; Seligmann 1995). The land reform was followed by the first massive rural-urban migration (Paerregaard 1997a, 1997b, 1998) and later by political conflict in the 1980s and early 1990s, which generated violence and poverty (Gorriti 1999; Mitchell 2006; Paerregaard 2002; Starn 1999). While neoliberalization and globalization created new social divides in the 1990s, they also provided the urban and rural working classes space for upward mobility (Cotler and Cuenca 2011; N. Lynch 2014). And after the Fujimori regime abruptly collapsed in 2001, Peru experienced an unprecedented period of stability, with five democratically elected governments in a row. More recently, however, the stability has been shaken by corruption charges, which have brought down several of the country's presidents, and by Peruvians' loss of trust in their politicians' integrity.

A number of developments have accelerated Peru's transformation in the first two decades of the twenty-first century. First, in the 2000s millions of Peruvians emigrated to a range of destinations across the world, draining the country of its human resources and creating a global diaspora.[5] In the 2010s the exodus slowed down and was replaced by other population movements, including a steadily growing flow of domestic and international tourists and, more recently, a wave of refugees from Venezuela and migrants from other countries in the Americas who seek haven in Peru or use the country as passage to other countries in South America.[6] Second, between 2002 and 2012 Peru experienced an economic boom that sparked growth not only in the cities but also in rural areas.[7] Third, the introduction of new communication technologies has connected people in Peru's remote areas with the outside world and made them part of the global flow of news and information (Ødegaard 2020; Stensrud 2017). At the same time, the country has reviewed its precolonial past, and even though racism and other forms of discrimination are still widespread in Peru, today a growing number of Peruvians recognize the country's indigenous cultures as an important part of its national heritage (Greene 2009). As a result, Andean dance, music, handicrafts, and food have become emblematic of Peruvian identity both inside and outside the country (Altamirano and Altamirano 2019; Nelson 2016; Paerregaard 2018c; Pérez Gálvez et al. 2017).

Peru's development in the past fifty years has transformed the country not only economically but also socially and culturally. Yet even though urbanization and modernization have generated prosperity and provided people with new life opportunities, they have also created environmental problems and feelings of insecurity, prompting many Peruvians to question the country's development model. This is driven by the mining industry, which produces 60 percent of Peru's exports. However, the industry also triggers social unrest and violent protests in the Andean highlands, where it is concentrated and where it contaminates the environment and pollutes the water (Li 2015; Ødegaard and Rivera Andía 2019). In the research for answers to the new challenges Peru is facing, a growing number of Peruvians are turning toward the country's cultural past and taking a renewed interest in Andean ideas about human-environment relations and the ritual practices that model them. Many Andean communities have abandoned these customs and traditions and their adoration of *apus* and pre-Hispanic *huakas* (and sometimes also pre-Inca ancestors known as *gentiles*), but in some places they still prevail, as I discuss in the two first case studies. Moreover, in some parts of the country mountains, glaciers, and ice caps are iconic sites of Andean-Christian pilgrimages that in recent years have attracted a growing number of urban Peruvians, a phenomenon I explore in the two last case studies. One of the forces behind this surge of interest in Andean culture and identity is climate change and the water crisis it creates in Peru.

PERU'S WATER CRISIS

The production of minerals, oil and gas, timber, and agricultural products that drives Peru's growth has generated thousands of jobs and prosperity for a large section of its urban and coastal population. However, it also generates discontent among the people who live in the vicinity of the production sites and who suffer from their environmental impacts (Triscritti 2013). Moreover, both mining and cash cropping are water-demanding activities that cause water stress and water conflicts, which the recurrent clashes between water users, mining companies, the agroindustry, and the state are evidence of (Gustavsson 2016; Isch, Boelens, and Peña 2012; Li 2015; Oré 2005; Urteaga and Boelens 2006). These frictions are aggravated by the big water infrastructures that the Peruvian state constructs, which transports water from the highlands to the coastal desert to meet the agroindustry's water needs (Andersen 2016; Paerregaard, Ullberg, and Brandshaug 2020; Poupeau and González 2010; Rasmussen 2016a; Stensrud 2016b, 2021; Ullberg 2019).

To mitigate Peru's many environmental issues and cope with the country's water crisis, the government introduced a new water law in 2009 that aims to implement a uniform system of water tariffs and taxes and establish a new institutional setting to negotiate water demands and water rights (ANA 2010; Oré et al. 2009; Roa-García, Urteaga-Crovetto, and Bustamente-Zenteno 2015).[8] The law affirms the definition of Peru's water resources in the previous water law from 1969 as the property of the state and a public good that cannot be privatized (Guillet 1992, 99–116). It ranks human needs for water higher than other demands and those of cattle breeding and agriculture higher than those of mining, the agroindustry, and so forth. To achieve its aim, the 2009 law invites the country's water users to negotiate water rights and manage water allocation in so-called water basin councils, an institutional setting designed by the World Bank's IWRM (Integrated Water Resource Management) approach (Allouche 2016; Andersen 2019; Tortajada 2015).[9] Compared to the water laws of other Andean countries, which emphasize equity and sustainability, however, efficiency prevails in Peru's 2009 law (Roa-García 2014), which recognizes water as a basic need but nevertheless requires that people obtain the state's permission to use it. Moreover, after obtaining the right to access water, the users must pay a tax (*retribución económica*) for their water consumption to the national water authority (Autoridad Nacional de Agua, ANA) as well as two different water fees (*tarifas para el uso de agua*): one to their water users' organization (Junta de Usuarios) for their use of the canals and reservoirs of their own communities and another to the regional water authority (Autoridad Local de Agua, ALA) for their use of the state's water infrastructures.[10] And while the law acknowledges the right of Peru's indigenous communities to use and value water according to their cultural traditions and customs, it does not exempt them from paying the water tax and the water tariff.[11]

But even though the 2009 law represents a milestone in the state's efforts to improve Peru's water governance, it highlights the state's absence in Peru's marginal areas and neglects the underlying social tensions that fuel Peru's water crisis (del Castillo 2011; Paerregaard, Stensrud, and Andersen 2016; Paerregaard, Ullberg, and Brandshaug 2020; Rasmussen 2016b, 2016c; Stensrud 2016c, 2019). Struggles over water rights, water values, and water accountability pervade the relations between Andean communities, the mining companies, agroindustry, and the Peruvian state, which makes water an issue of economic and political contestation and social and cultural dispute (Boelens 2015; Stensrud 2021). Discontent with the tax and the tariff is amplified by Andean water users' mistrust of the state and other external agents, such as big landowners and mining companies, and taps into a century-long conflict over water rights in the Andes that has made water struggles a critical emblem of the communities' culture and identity and a highly charged political question in Peru. In the words of Rudgerd Boelens: "The struggle over water rights is simultaneously a battle over resources and legitimacy: the legitimacy to formulate and enforce water rights and to *exist* as water user collectivities, to have sufficient control over one's own future" (2008, 50).[12] To the communities, therefore, the tax and tariff of the 2009 law are not merely payments for their water use and the maintenance of their water canals and reservoirs. As the book's first case study shows, they imply the communities' recognition of the state as the legitimate owner of their water resources and the user organizations as the proper authority to manage their water supply and water infrastructures and to represent their interests in the state's water institutions. To cede these powers to the state and user organizations, the communities demand tangible evidence of their ability to enhance their water supply and improve their water management, as illustrated in the book's second case study.

Other important measures taken by the Peruvian government to address the country's water crisis include the national strategy to mitigate the impact of climate change (ENCC 2015), which was introduced in 2015.[13] Following the principles of UNFCCC (the United Nations Framework Convention on Climate Change), Estratégia Nacional ante el Cambio Climático establishes the nature of Peru's climate vulnerability and defines the areas of intervention the Peruvian government needs to act in to reduce the emission of the country's greenhouse gasses and adapt it to a future with changing temperatures, irregular rainfall, water shortages, and other forms of environmental change.[14] On paper the climate change strategy sounds promising. It names some of Peru's most vulnerable populations (e.g., indigenous people in the highlands and the Amazon) and defines climate vulnerability in terms of inequality and poverty (Caine 2021). It also suggests improvement of the water infrastructure and preparing agriculture for plant diseases and other climate-related impacts among the rural poor and recommends their participation in what is called integrated risk management to achieve the declared adaptation goals and the planned reduction in vulnerability. However, as discussed regarding the 2009 water law, involving Peru's rural communities in national policies and giving them a political voice, as ENCC's bottom-up approach suggests, pose huge challenges because of the mutual mistrust between governmental institutions and marginal populations. So far it has had no bearing on the four case studies explored in this book. But even though the promises of Peru's national strategy to mitigate climate change are long in coming, other government actions to protect the country's environment are showing results. As the book's two last case studies show, Peru's national and regional governments have restricted people's access to the country's retreating glaciers and set up rules for the activities that pilgrims, tourists, and others can engage in when visiting them. I discuss the consequences of these measures in chapters 4 and 5.

REPLICATING METABOLISM

Contemporary ethnographies show that mountains continue to play a critical role in the lifeworld of Andean people (de la Cadena 2015). Some communities view them as emblems of their ethnic descent and identity (Gelles 2000), while others ascribe them human attributes and read them into a universe of kin and gender relationships (Salomon 2018). In a similar vein, anthropologists have demonstrated how Andean people reimagine mountains in new geographical settings when they migrate to the cities (Fine-Dare 2019; Ødegaard 2011). Supporting the classical literature on sacrifice, which argued that sacrifice was a means of communication with the sacred (Hubert and Mauss 1964), ethnographic research also shows that people perceive mountains and other more-than-human agents as powerful and possibly dangerous forces that may do harm to humans (Gose 1994; Isbell 1978, 151–163; Paerregaard 1989, 1997a, 206–211). These studies describe how people socialize and establish a relationship of exchange with the mountains to mitigate the danger they represent and to ask for personal favors and help cure people who fall ill or otherwise are in need. They also suggest that in return for their offering gifts, the communities expect the mountains to provide them with water, food, and other material goods; ensure the fertility of both humans and animals; and produce a good harvest (Allen 1988; Cometti 2020a, 2020b; Salas Carreño 2019). In the four case studies I elaborate on this topic, discussing how humans' cohabitation with mountains implies an element of safeguarding as well as threat, an ambiguity Andean people sometimes phrase in terms of who consumes whom: either humans make offerings to the mountains, which in return allow them to "eat" their resources, or the mountains go hungry and therefore "eat" humans, a phrasing that is particularly common in regions with large-scale mining (Nash 1979; Salas Carreño 2017).

Ethnographic accounts also suggest that the syncretic cosmology of Andean and Catholic beliefs that emerged after the conquest still informs the water management of some Andean communities and that they continue to perceive mountains as masters of the water flow and appease them with offering gifts (Bastien 1978; Isbell 1978; Paerregaard 2013a; Stensrud 2016a). Moreover, scholarly works report that some Andean communities regard water as a vital substance of not only material but also symbolic value (Andolina 2012; Gose 1986; Nash 1979; Taussig 1980). Other studies suggest that water's aptness to unite and connect as well as dissolve and disconnect is read as proof that it is a living matter (Paerregaard 2018b; Salomon 2018, 28) and that it possesses the ability to think, learn, and respond to human contact (Brandshaug 2019, 2021; Stensrud 2021; Treacy 1994, 113–140; Valderrama and Escalante 1988).[15]

Several studies also reveal that water infrastructure and management constitute a fundamental principle for the political organization and cosmological structure of many communities that give irrigation canals names and associate them with their social histories (Guillet 1992; Mitchell and Guillet 1994; Paerregaard 1997a, 2014b; Rasmussen 2015). These insights suggest that water's physical characteristics induce Andean people to associate it with transcendental and metaphysical aspects of life and to coin water rights as a question of not only legal or human rights but also cultural values and ethics. As David Groenfelt points out: "Values and ethics pervade water governance both through decisions about the governance regime itself (values about governance in general which are applied to water governance as well as other forms of governance), and through decisions about how water should be used (values about water)" (2013, 107).[16] But if water values transcend water governance and water justice in the Andes (as well as in other parts of the world), how do they come into play in Andean offerings, and how do the Andean people construe these as a means to access the water? And more specifically, how do Andean people imagine the offering items and the ritual in which these are transformed into a gift as an enactment of the hydrological cycle?

My suggestion is that as a metabolic catalyst, Andean offerings serve as not only a tool to communicate with the sacred (Hubert and Mauss 1964) but also a regulator of humans' use of natural resources, similarly to the Kaiko ritual. More specifically, offerings are a way of adapting to an environment of scarce resources and reinforcing community resilience in a time of rapid environmental change.[17] However, while I propose that Andean offerings seek to restore a symbiotic human/nonhuman relationship in the same way that Kaiko does, my analysis of the social dynamics that drive the ritual differs from Rappaport's study. In Rappaport's account, Kaiko constitutes a self-contained cultural practice. My analysis, by contrast, emphasizes the long history of social contestation that has shaped modern-day Andean offerings. And while Rappaport presents Kaiko as a ritual ingrained in a culture that appears isolated and static, I examine Andean offerings as a practice that has been made and remade to adapt to the changing social and political world. Hence, rather than evolving into a maladaptive construct, as Rappaport predicted could happen to culture, I view Andean rituals and the cosmology that fuels them as both resilient and malleable.

As pointed out earlier, Andean people have continuously adjusted their offerings to the surrounding world. Still, these comprise a repertoire of practices that is of central importance to the replication of water metabolism and the focal point of the four case studies. One part of this repertoire is related to the materiality of the offering items, another to the ritual's social configuration, and a third to its ideas of accountability and viability. Rehearsing the metabolic process that takes place in nature, the offerings' participants trigger a "change of matter" of a collection of items by setting fire to them, throwing liquids in the air, chewing coca leaves, drinking spirits, and smoking cigarettes. The metabolized objects represent key elements in both the physical and cosmological worlds that Andean people inhabit and that constitute the environment that encircles their freshwater resources. The items vary from one community to another, but often they include organic elements from the ocean (seawater, seaweed, and shellfish) and the fields and mountains (different classes of corn, quinoa, and wild herbs); metal pieces from underground (gold and silver); living animals (guinea pigs, sheep and llamas) or animal fetuses and grease; and spiritual items such as corn beer, wine, spirits, cigarettes, and coca leaves (Bolin 1998, 33–43; Salomon 2018, 63–67; Stensrud 2021; Valderrama and Escalante 1988, 109–123).[18] Importantly, the ritual transforms not only the offering items' chemical composition, as these are burned, consumed, and disposed, but also their symbolic value, as their destruction and decomposition recalibrate them into a gift.

Even though the mountains' meltwater is not itself metabolized in the ritual, the offering items include liquids—wine, spirits, and corn beer—that undergo a "change of matter" and therefore are metabolized as they are consumed by the participants and become bodily liquids, or as they are dropped to the ground or left (in bottles or other containers) at the offering site, where they are eventually absorbed by other materials. Whether liquid or solid, the offering items are critical to the water metabolism the ritual sets in motion, as it is their "change of matter" that produces the gift, an idea prevailing in many other Andean rituals such as offerings to *pachamama* (Quechua: Mother Earth) to yield a good harvest or to *gentiles* (Quechua: Andean people's distant ancestors) to cure people who fall ill (Paerregaard 1987a, 1989). Smoothing the water flow by making offerings to the mountains is therefore just one of many examples of how Andean people establish a mutual and viable relationship with the environment by enacting the "change of matter" of nature's elements and reassembling them as a gift to the powers they believe control their lives.[19]

But it's not only the offering practice and its ceremonial repertoire that demonstrate persistence. Even though Andean people have remodeled the cosmology they inherited from the Incas and have altered the human figures of its pantheon several times since the conquest, contemporary ethnographic research suggests that the structure of its anthropomorphic representation has changed little since the early republican era. The very idea that mountains are human comes to bear in Joseph Bastien's ethnography of the Kaata community in northwestern highland Bolivia. Bastien writes: "Legends and rituals symbolize Mount Kaata as a human mountain. Kaatans name the places of the mountain according to the anatomy of the human body" (1978, 37). The Kaatans' anatomical paradigm resonates with ethnographic accounts from other parts of the Andes. From his study of the community of Rapaz, Frank Salomon learned that "mountains have feelings much like those of humans, including possessiveness and anger. Mountains' sense of right and wrong is a moral-political system analogous to human community" (2018, 612). Salomon also observed that Rapaz ranks nearby mountains according to their heights and importance as water suppliers and uses political titles to address their deities. The mountains' social order is a mirror of the community's own hierarchy of political leaders and administrative offices as well as the national government's division of political powers, with the president as the highest ranked. Salomon recounts: "By a further analogy, mountains of a given landscape are political beings; though rivalrous and unequal, they work together as a 'council' and stand to each other in civic hierarchy as 'president' (the highest mountain) and other ranks" (2018, 612).

The idea of mountains as icons of Peru's political system also prevails in Paul Gelles's ethnography of the community of Cabanaconde, which I elaborate on in chapter 3. In Cabanaconde, mountains are known as *cabildos* (Spanish: town councils), which "are the mountains that have authority over the town of Cabanaconde; they are the power figures to whom the Cabañenos owe allegiance and respect and to whom they must make sacrifices and pay tribute. In return, the cabildos watch over and protect the Cabañenos" (Gelles 2000, 83). In a similar vein, anthropologists have shown that the mountains' social status reflects Peru's racial and ethnic divides. In the community of Chuschi, where mountains are called *wamanis*, Billie Jean Isbell found that "the Wamanis preside over territories and have an organizational hierarchy likened to provincial governmental structure. They are described as tall, white, bearded males who dress elaborately in western dress" (1978, 59). By the same token, Peter Gose reports from the community of Huaquirca that "when the mountain deities take human form, it is stereotypically as large, blonde, blued-eyed men, dressed in the fancy clothes and riding boots that would have been worn by a *hacendado* [estate owner] in the earlier part of this century" (1994, 209). These findings resonate with Juan Ansión's study in Ayacucho. He writes: "The president [of Peru] is perceived as a monarch, or as an Inca. Opposite him rises the local power of the Wamani, who definitely is inferior (he is a second God after Jesus Christ), but who prodigiously sees after the local problems if you respect and venerate him. And as the power that currently dominates in the country is the one of white people, the Wamani is also attributed the features of white people: color, hair, etc." (1987, 189). In other words, Andean people's configuration of mountains as images of Peru's political rulers and ethnic elite that emerged after the country gained independence in 1821 is still vivid.

A third aspect of Andean offering practice that needs attention is the semantics of its terminology and the ideas of reciprocity, accountability, and viability it glosses over. In some parts of the Andes the tribute is known as *t'inka* (Paerregaard 1989; Valderrama and Escalante 1988), which according to Gose means libation in Quechua (Gose 1994). Gose reports that the name "applies to the sprinkling of alcohol on the ground or wafting of its vapours towards the mountains" (1994, 298).[20] In other places it is called *pagachu*, which is a Quechua term derived from the Spanish word *pago* meaning "payment" and from the Quechua word *apu* meaning "lord" that is shortened to *pu*. As *pagachu*, the tribute is a payment in a contractual relationship between humans and the mountains who control the water and demand gifts in exchange for its future delivery (Gose 1994; Isbell 1978; Molinié 2019; Paerregaard 1994a; Salomon 2018). In support of this interpretation, Salomon writes that in the community of Rapaz the offering is known not only as *pago* but also as *derecho* (Spanish: right) or *cumplimiento* (Spanish: fulfillment), which means that "something is owed to the mountain" (2018, 67). At the same time, Salomon explains that "when humans perform *qarakuna* [the act of serving the mountains with goods] the mountains eat, smoke, take coca, experience pleasant sociability, and thereby incur a debt to people," which suggests that the relationship between humans and nonhuman agents is mutual. Whereas humans fulfill an obligation to the nonhuman powers when making the offering, the latter put themselves in debt to the former by receiving it.

The reciprocity, however, is fragile. As the terms *pago*, *pagachu*, and *derecho* reveal, the offering is a payment, which the mountains have the right to claim. And as is evident from several of the ethnographic accounts referred to earlier, and as I show in the first case study, the mountains have the power to not only hold back the water but also punish humans who show lack of respect for their favors or fail to pay them tribute. Even though the receivers reciprocate the offering gift with water, then, they hold the givers accountable for managing it correctly and providing more gifts. Recalling Rappaport's claim that rituals regulate the human-nature nexus, offerings can thus be viewed as a mechanism to ensure a sustainable use of a good in chronic shortage, which I argue is pivotal to our understanding of how Andean people adapt to climate change and which is central in the analysis of my four case studies.

CONTEXTUALIZING THE FIELD SITES

As neighboring districts at the bottom of the Colca Valley in Peru's southern highlands, Tapay and Cabanaconde have much in common.[21] The inhabitants of both districts were both subject to the policy of *reducciones*, and even though Tapay is smaller demographically and more isolated geographically than Cabanaconde, their economic and political development differs little. Thus a massive increase in tourism, national as well as international, in the past two decades in the Colca Valley has given rise to new livelihoods and a flow of information from the outside world to both Tapay and Cabanaconde. Finally, Peru's recent economic boom has left its mark on both districts, where the national and regional governments have built new schools, health clinics, roads, irrigation canals and reservoirs, and infrastructure to provide individual households with drinking water, electricity, and sewage. However, the development of Tapay and Cabanaconde differs in several ways. Since the 1960s both districts have experienced extensive out-migration that has drained them of their able-bodied members. But while Tapay's migration flow is almost entirely directed toward Peru's major cities, in Cabanaconde it is not only rural-urban but also transnational. The district now has migrant colonies in the United States, Spain, and Chile that have considerable leverage in its internal affairs.

Another important difference in Tapay's and Cabanaconde's development is their water supply and management and the capacity to adapt to climate change. In 1983 the Peruvian government completed the construction of the Majes channel, which transports water from the Colca River to an area called Majes Plains, where it is used to irrigate the coastal desert. To access water from the channel that runs through their territory, the villagers made a hole in it, which initially triggered a clash with the national police but later prompted the administrators of the channel to acknowledge their claim to water. As a result, Cabanaconde doubled its irrigated land base, enabling it to enhance agricultural production and invest in livestock farming. Today, Cabanaconde receives most of its water from the Majes channel, which has replaced its former water source, Mount Hualca Hualca (6,025 m). Due to rising temperatures and rapid snow melt, Tapay, on the other hand, has experienced a continuous shrinking of its water resources in the past three decades, and today the district only cultivates one-third of its agricultural land. The opening of a gold mine in its territory that not only takes water from the same water source as Tapay but also represents a serious threat of contamination to its environment has deepened the villagers' concern for their future. To gain support, the mine has offered Tapay's villagers and migrants employment and promised to build a channel to direct water to the district from a nearby river, which in the near future may replace its principal water source, Mount Seprigina (5,450 m). The two districts' engagement with external agents and the different kinds of support these offer Tapay and Cabanaconde have important implications for not only their possibilities of adapting to climate change and reorganizing their water management but also their perceptions of political and religious power and relations with the state and the mountain deities. Thus, while Cabanaconde has adapted a state model to manage its irrigation system and ceased making offerings to Hualca Hualca, Tapay still employs a community model to irrigate and continues to pay tribute to Seprigina. By the same token, Cabanaconde pays the water tax and water tariff required by the 2009 water law, while many in Tapay refuse to accept the state's right to charge them for their water use.[22]

Located respectively in Peru's central and southeastern highlands, Mount Qulque Punku, its neighboring Mount Ausungate (6,372 m), and Mount Huaytapallana (5,557 m) are the scenes of annual religious gatherings that I have attended several times between 2014 and 2017. They follow a century-long tradition of regional pilgrimages in the Andes that honor and adore its mountains and glaciers. The history, scope, and structure of the two pilgrimages, however, are quite different. The pilgrimage of Quyllurit'i, as the adoration of Qulque Punku is known, can be traced back to Peru's colonial period and attracts tens of thousands of pilgrims from not only the region of Cusco but also other parts of Peru. Moreover, it taps into a mixture of belief systems that comprises Christian and Andean religious elements as well as Evangelist and Andean commercial elements, practiced in a variety of ways including the adoration of the Catholic saint of Quyllurit'i; the honoring of Qulque Punku's glacier; the performance of a multitude of dance and music groups; and the sale of food, camping equipment, and *alasitas* (i.e., miniature figures or toy models of items and phony copies of entitlements and currencies that buyers desire; see Golte and León Gabriel 2014). In contrast, the pilgrimage of Huaytapallana began less than thirty years ago and is attended by around one thousand pilgrims who mostly come from the nearby city of Huancayo and its surroundings. And unlike the pilgrimage of Quyllurit'i, it only draws on ideas and symbols from indigenous metaphysics. Known as *la cosmovisión andina* (Spanish: the Andean cosmovision, a synonym for the pre-Iberian cosmology of Andean people), it presents the universe as a flow of energy that connects all its elements, including humans, mountains, rivers, the wind, the sun, the moon, and the rest of the cosmos. From this perspective, humans must seek to live in balance with the nonhuman realm, which the pilgrims do by honoring and paying tribute to Huaytapallana.

Notwithstanding these differences, the pilgrims of both gatherings are moved by a concern for Peru's water scarcity, in particular the retreat of Qulque Punku's and Huaytapallana's glaciers. The growing interest in the mountains has also led to an increasing attendance of the two pilgrimages and, paradoxically, an awareness of the environmental threat the ritual traditions and the pilgrims' physical presence represent to them. Quyllurit'i and Huaytapallana have also drawn the attention of political actors who are trying to regulate the pilgrims' activities and contain their impact on the mountains. In 2011 the United Nations Educational, Scientific, and Cultural Organization (UNESCO) declared Quyllurit'i an intangible heritage site, and subsequently Peru's Ministry of Culture has overseen its cultural preservation. In the same year the Peruvian government declared Huaytapallana a conservation area and authorized the regional government of Junín to protect its environment. Even though the mandates of the two institutions vary, one preserving Quyllurit'i cultural tradition and the other conserving Huaytapallana's environment, they both submit the pilgrimages to administrative control and compel their organizers to cooperate with Peru's national and regional authorities. In the wake of the dramatic glacier retreats of Qulque Punku and Huaytapallana and the recent political interference in the organization of the pilgrimages, a new understanding of how anthropogenic climate change jeopardizes the planet's future and upsets humans' relations to other forms of life is now emerging. In a similar vein, many pilgrims are disturbed by the amount of trash and other human remains that the pilgrimages produce, and while some have started to question the meaning of their ritual practices, others worry whether they are contributing to the mountains' death.

Tapay

THE OFFERING MUST GO ON

"Get up. It's already late," I heard Godofredo saying outside my door. I rushed out of my sleeping bag, got dressed, and hurried over to the kitchen, where Godofredo and three others were eating breakfast, which his wife had prepared. It was five in the morning on All Saints' Day (November 1), and we had a long and strenuous day ahead of us. In 2011 Godofredo was *regidor* (Spanish: water allocator) in Tapay's main hamlet, and following the community's traditions, he had assumed the task of organizing the annual offering to Seprigina, Tapay's highest mountain and principal water source. I had known Godofredo since I did my first fieldwork in Tapay thirty-five years ago, and he had invited me to accompany him. Godofredo had also contracted three villagers to help him conduct the offering ceremony and to lead the four mules he had hired for the occasion. One of the men was a *paqu*, a specialist in making the *pago*. The man had prepared for the trip the night before by arranging an offering ceremony in Godofredo's house to ask Seprigina for permission to visit it. And as the mountain had accepted his petition, the prospects of our mission looked good.

Even though I had passed by Seprigina several times when traveling to Tapay's *puna*, the tundra-like grassland above 4,000 meters, during my previous stays in the district, it was my first visit to the mountain's summit. After two hours' mule ride and three hours' climbing, we reached our goal: a spring on the side of the summit that the villagers regard as their principal water source. The sky was blue, and even though the air was cool the sun was strong enough to warm us up. While the rest of us chewed coca leaves, drank wine and spirits, and smoked cigarettes, the *pagu* prepared the *iranta*, a collection of offering items, which he placed under a rock that covered the hole to the spring. Meanwhile Godofredo burned incense on a stone that he lifted into the air. With a low voice he started to talk with the mountain, asking it to receive the gift we were about to bring and, in return, bless the community with plenty of water. Godofredo then passed the stone with the burning incense on to his assistants and me, and we all replicated his gestures, expressing our respect for the mountain and its power.

On the trip back we followed the creek that runs from the spring down to Tapay, which is a more direct but also steeper and more arduous route. At the takeoff where Tapay's main canal taps water from the creek, Godofredo's wife and a group of villagers were waiting with chicha, spirits, and beer, which we drank while the *pagu* repeated the offering he had previously made at the summit. According to Godofredo, the gathering was part of his commitment as *regidor*. Making offerings to Seprigina's summit smooths the circulation of the mountain's water, while paying tribute to the offtake facilitates its flow from the creek into Tapay's water infrastructure. The party broke up at dusk when we all embarked on the last descent to the hamlet, which we reached an hour later, exhilarated but exhausted.

As the first of this book's four ethnographic case studies, Tapay illustrates how the ritualization of water metabolism works in a situation where the community that performs the offering entirely relies on the water flow it is intended to smooth. Using the water of Seprigina to both meet domestic needs and irrigate the fields, Tapay illustrates what I call existential metabolism, a ritual enactment of the hydrological cycle that is embedded in the community's daily life and local cosmology.

INSIDERS AND OUTSIDERS

Godofredo's organization of the annual offering to Segrigina may seem paradoxical considering his migration history and long absence from Tapay. Why assume the office as *regidor* if you do not make a living as an agriculturalist in the district, and why commit yourself to a ritual duty that is detached from other activities in your life? Like many of his fellow villagers, Godofredo had left Tapay when he was young, and he had only returned after retiring.

In fact, Godofredo's migration story reminds one of those of other villagers, except that the bulk of Tapay's migrants never return. For many years, staying put in the district has not been regarded as a viable option for young villagers, who leave Tapay before they come of age to *progesar* (Spanish: to make progress)—that is, find work, make money, and establish a family in

FIGURE 1. The district of Tapay, in the Colca Valley, Peru, 2011. Photo by author.

Peru's cities. As a result, Tapay has experienced a notable demographic decline in the past five decades, and today the district is mostly inhabited by either very young or old people. Its migrant population, on the other hand, views it as a place to seek distraction from city life, visit relatives, or, as Godofredo has done, enjoy retirement. Even so, from time to time the office of *regidor* is filled by return or visiting migrants who affirm their loyalty and attachment to Tapay by making the strenuous trip to Seprigina's summit on November 1 and paying tribute to the mountain.

The population's division into insiders and outsiders is a consequence of the district's physical isolation and lack of opportunities. Located between 2,300 and 5,400 meters in the lower end of the Colca Valley, Tapay lies on the fringes of one of the deepest canyons in the world and is surrounded by mountains (see figure 1).

When I arrived at Tapay in 1986 it could only be reached on foot or by mule, and the district had to wait until 2016 for the regional government to build a single-track dirt road, which now connects it to neighboring Caba-naconde. Also, during most of my fieldwork I have cooked my food on open fire and used candles to light the house I rented, and it was only in 2009 that electricity was installed in Tapay. Other public services have been added more recently. However, it would be a mistake to assume that the villagers live in a timeless Andean past. A brief look into its colonial and postcolonial history and demographic development shows that Tapay has been the target of recurrent in-and out-migration and that the district's population movements and the division into insiders and outsider is a generic trait of its social organization.

The population movement that Tapay has experienced in the past two centuries is reflected in its demographic development and size compared with other Colca districts.[1] One of the main reasons for Tapay's demographic growth was an influx of outsiders searching for opportunities to exploit not only the land but also local labor. In the mid-nineteenth century a mestizo priest from Arequipa settled in Tapay, bringing along his brother and sister-in-law, who had four sons and one daughter. Even though the family had no relatives or properties in Tapay, the sons became the biggest landowners in the district at the turn of the century. Educated by Spanish-speaking teachers but married to local women, they used their bilingual proficiency in Spanish and Quechua and familiarity with both the Andean and mestizo worlds to occupy the district's political and administrative offices and establish themselves as *gamonales*. At that time the posts of *alcalde* (mayor), *gobernador* (the government's administrative representative), and *juez de paz* (the peace judge) were *inmoviles* (Spanish: permanent), and over the years the brothers used them not only to command the villagers to do corvée labor but also to grab their land.[2] To legitimize their power the brothers portrayed themselves as educated and Spanish-speaking *mistis* (see chapter 1), in contrast to the Quechua-speaking population, who had little or no knowledge of Peru's urban and coastal Spanish-speaking world and therefore were classified as *indios* (Spanish: Indians). The *misti* brothers displayed their ethnic status in various ways, including their outfits (wearing shoes instead of sandals), religious customs (celebrating Catholic saints instead of mountain deities), and work habits (using an oxen plow instead of a foot plow). Even though most of the brothers' children integrated into Tapay's native population, a few of them continued to claim the family's ethnic status as *mistis*, and as its biggest landowners, they have dominated life in the district up to recently.

Tapay's demographic development took a dramatic turn in the second half of the twentieth century when the population dropped significantly.[3] The decline continued in the first decade of the twenty-first century, reaching a record low of 671 in 2007 (INEI 2017). The main cause of the decline was massive out-migration. The first villagers to leave were men who took seasonal work on the coastal plains of Majes to make an income in cash. Others were drafted by the Peruvian army and spent several years in military barracks on Peru's southern coast. The army submitted them to harsh discipline, but it also taught them to read and write Spanish and opened their eyes to the world outside Tapay. The experience enticed the young men to challenge the *misti* family's dominance upon their return to the district. Up to the 1950s schooling was considered a privilege reserved for Tapay's Spanish-speaking population, and the *misti* family did everything it could to keep the children of Quechua-speaking villagers away from the district's school. Several senior villagers have told me that, paradoxically, the *misti* family would hide not only their own sons but also the rest of the district's young men when the military made roundups in Tapay to draft them. Remembering the days of *misti* dominance, a villager recounted: "The *mistis* would do anything to prevent us from traveling, learning Spanish, and getting to know the world outside Tapay."

Another senior villager recalled how he confronted one of the *misti* brothers' sons after completing his military service in 1942. Before returning to Tapay, the man followed the suggestion of one of his army superiors to change the ethnic label on his birth certificate from *indígena* (Spanish: indigenous) to *trigueño* (Spanish: mixed race). The man told me that it was one of the *misti* brother's sons who had filled out his birth certificate and written *indígena*, which had made him feel "as [if] I was from some remote place in the jungle" and like "a racial inferior Peruvian." He said: "The officer's suggestion to change my birth certificate made me think: 'Why not write *español*?'" referring to the label the *mistis* used for themselves, which what was he did. Learning that the man had changed his certificate, the *mistis* got upset, but as other young villagers followed his example and contested their status as superior villagers, the *misti* family's power base started to crumble. And as out-migration gained momentum in the second half of the twentieth century, the *misti* brothers' descendants joined the steady stream of villagers who left Tapay in the search for a better future in the city. Today, it is only when its members sponsor Tapay's five-day-long fiesta in February that the *misti* family members for a short while reclaim the privileged status of their late kinsmen.

Just as immigration was a strategy for outsiders to establish a power position as *mistis* in Tapay and exploit its manpower and resources in the nineteenth century, out-migration has been a way for the villagers to break out of the rural world that had confined their parents, grandparents, and great-grandparents to life as *indio* and *indígena*. The pull factor of this population movement was the growing wealth in Peru's cities and the dream of *progresar*. The push factor, on the other hand, was Tapay's physical isolation and lack of development, which has generated a feeling of being left behind and a sense of abandonment (Rasmussen 2017).[4] The out-migration has not only drained Tapay of its able-bodied population but also changed its gender and age composition. While women slightly outnumbered men thirty-five years ago when I did my first fieldwork in the district, men constituted the bulk of the population in 2017.[5] Similarly, villagers age sixty-five or older now make up more than one-quarter of the population.[6]

The development is notable. Tapay's population is both decreasing and aging, which has repercussions for its possibilities to adapt to climate change and manage its natural resources including the shrinking water supply, and as the vignette introducing this chapter illustrates, for its persistent tradition of paying tribute to Seprigina, which is practiced by villagers as well as migrants.[7] To understand the meaning of Tapay's offering practices, however, a description of its ecology, water resources, and irrigation system is required.

TAMING WATER

Tapay's population is not only divided into insiders and outsiders. It is also dispersed geographically, which has implications for the district's management of natural resources such as water. Unlike the populations of other Colca districts that are concentrated in one major and a few smaller settlements, all villagers of Tapay live in small hamlets.[8] The bulk of the population is scattered in nine hamlets located below 4,000 meters, while a small group of people live in clusters of houses on the *puna*. At a district level, the population is divided into moieties: Hanansaya, which apart from being the district's main hamlet (called Tapay) comprises three settlements, and the *puna* population and Urinsaya, which includes the rest of the hamlets. In the valley the principal livelihood is agriculture, which yields corn, potatoes, beans, and other crops and provides the villagers with their basic foodstuffs and fruit (apples, pears, peaches, figs, etc.), which they exchange for other produce and items on the regional barter market (Paerregaard 1992). Another important product is cochineal, a louse that lives on a cactus known by the same name and that when dried is used as dye in cosmetics and food. The global demand for natural dye has occasionally led to a boom in the trade of cochineal in Peru, providing the villagers with a substantial cash income (Paerregaard 1997a, 2000). On the *puna*, the herders raise llamas, alpacas, and, to a smaller extent, cattle and make a living selling fresh and dried meat or exchanging it for other products. When I conducted my first fieldwork in 1986, the villagers engaged in a regional barter market in which they exchanged fruit for corn, grain, and commercial products such as oil, matches, candles, and soap, but today they buy most of these items in the shops of Cabanaconde and the provincial capital of Chivay (Paerregaard 1993).

Except for cochineal, all crops and fruits in Tapay depend on irrigation. In most Colca villages this is supplied by a few main water sources: rivers, streams, main canals, or water springs.[9] Tapay's water supply, however, comprises fifty-three water sources, of which twenty-one are *puquios* (Spanish: springs) and thirty-two are *tomas* (Spanish: offtakes) from six rivers and streams that carry the melted snow from the surrounding mountain peaks to the Colca River (Paerregaard 1994a). A total of twenty-one reservoirs in the district's many settlements allows the water users to store the meltwater overnight. The dispersed nature of Tapay's water supply is reflected in the organization of the district's irrigation system, which is divided into clusters delimited by the river, the offtake, or the spring that supply them with water as well as the reservoirs that store this and the canals that lead it to the fields (Guillet 1992).

Before the 2009 water law, Tapay largely maintained its water infrastructure and managed its water resources autonomously. The district was organized in informal water groups that appointed their own *regidores* and a separate *juez de agua* (Spanish: water judge) at communal assemblies (Paerregaard 1994a), and even though the state tried to enforce its control on several occasions, the villagers opposed it. During my fieldwork in 1986 I witnessed a visit by the Ministry of Agriculture to institute the water tariff, which had been introduced by the 1969 water law. The engineers from the ministry told the villagers that the law allowed Tapay's authorities to keep 90 percent of the tariff and that only 10 percent would be retained by the Junta Directiva de Regantes (now called Junta de Usuarios) located in the provincial capital Chivay. The villagers were familiar with the idea of paying tax (they had paid tax on land for several years), but many refused to pay the tariff. Resistance against forming *comisiones de riego* (Spanish: irrigation commissions), another requirement of the 1969 water law, was also widespread, and it was only in the mid-1990s that Tapay's water users started to form and join them.[10] Apart from abandoning the *juez de agua* office, however, the commissions continued to manage water as before.

When I revisited Tapay in 2015 the users of the district's main irrigation clusters had formed *comités de usuarios de agua* (Spanish: water user committees), as stipulated by the 2009 water law, in all hamlets.[11] Hence, four water committees are now in charge of managing the water coming from the district's three principal rivers (Seprigina, Molloco, and Tampoña) and its biggest spring, Oqto. They also operate and maintain the canals and the reservoirs that transport and store water from these sources. Despite the formalization and institutional change of the districts' water governance, though, the users continue to allocate water as they did when I first arrived. Each committee or water group appoints two *regidores*: a *hatun regidor* (Quechua/Spanish: big water allocator), who allocates water during the dry season from July and December, and a *huch'uy regidor* (Quechua/Spanish: little water allocator), who allocates water during the wet season from January to April. *Hatun regidor* is one of the most demanding public offices in Tapay. It is a mandatory task that passes in turn among the water users, and shunning it may lead to the loss of one's water rights. As water allocator it is the *regidor's* job to transport the water to the individual fields by opening and closing the offtakes, reservoirs, and canals.

Regidores may employ different methods to direct water to the fields. The most common method is *corte*, which involves allocating uphill from the lowest to the highest located fields and is used in Tapay's largest irrigation clusters, which are fed by offtakes from rivers or take water from the big reservoirs. Alternatively, the *brinco* method, in which the individual water users call on the *regidor* to allocate water to all their fields in a single round, is used in smaller irrigation clusters.[12] This method is increasingly gaining ground as more villagers are shifting from traditional crops such as corn to vegetables and commercial products such as avocado, which need irrigation on a more regular but not uniform basis. At the end of his/her term, the *hatun regidor* organizes the *yarqa aspiy* (Quechua: ditch cleaning), a compulsive workday for all water users, who come together not only to clean the canals and reservoirs but also to celebrate the outgoing *regidor*, who is expected to cover all costs for music, food, and beer (see figure 2).

To ensure smooth and fair water management, the user committee oversees the *regidor's* work; deals with cases of water theft; and resolves water disputes, which mostly occur because the users arrive late to receive water. During the months of irrigation, the water user's major challenge is to

FIGURE 2. Celebration of the regidor at *yaqar aspiy* in
Tapay, in the Colca Valley, Peru, 1986. Photo by author.

anticipate the *regidor*'s time schedule and be present when she/he opens a
boquerón (Spanish: big mouth), which is done by blocking the water flow
in the canal and digging a hole into the field. Once this gets flooded, it
is the owner's own responsibility to guide the water. Most water disputes
arise because of timing. Many villagers report that they often spend hours
waiting to receive water because others arrive late and therefore uphold
the *regidor*. To prevent such delays, rules concerning water distribution are
strict. Water users who are absent when the *regidor* appears may miss their
turn.[13] Depending on the crop and the irrigation cluster, such a sanction
can have serious consequences for the water user, who may have to wait a
month or more until she/he receives water. Ultimately, this may result in
losing the entire crop.

According to the 2009 law, the user committees are responsible for collecting both the water tax and water tariff from their members and handing the money over to ANA and the Junta de Usuarios in Chivay. The committees are also expected to collect the tax and the tariff from the users of Tapay's many water springs.[14] However, many water users evade the tax and the tariff, and the collection is lower than in other Colca districts. According to an engineer working in ALA's office in Pedregal whom I interviewed in 2011, only 70 percent of Tapay's water users pay the water tax. In 2015 the presidents of the user committees in Tapay's two largest settlements told similar stories. The first said that fewer than fifty of the main settlement's eighty-five water users had paid the water tariff that year, while the second said that fewer than forty out of sixty-three water users of the Cosñihuar-Malata settlement had paid it. The latter also informed me that the number of tariff-paying water users in the district's other settlements was close to zero (Paerregaard 2019a). The extensive tariff evasion means that Tapay's water committees only collect a fraction of the money they are required to send to the Junta de Usuarios; consequently, the latter provides them with little or no information and support. The collection of water taxes and water tariffs is complicated by lack of information on Tapay's water users. The Ministry of Agriculture has not yet conducted a PROFODUA in Tapay, as in the rest of the Colca region, and can therefore not provide the data on the villagers' landholdings needed to create an official register of the district's water users.[15] Consequently, Tapay's user committees still use their own incomplete and outdated member lists to collect the water tax, which allows many villagers to evade it.

The bulk of Tapay's water tax and tariff evaders are migrants who live in the cities of Lima and Arequipa and have either left their fields to others or abandoned them altogether. However, the evaders also include villagers who cultivate and irrigate their fields. Many of them say that the reason for their neglect is that they lack water.[16] When I interviewed the president of the user committee of Tapay's main hamlet in 2016 he affirmed this claim. "The water supply is less than half of what it was thirty years ago," the man said. Moreover, the water tax and the water tariff are opposed by villagers, who claim that Tapay's water resources are a gift bestowed by the mountain deities and that the water infrastructure is a heritage of their forefathers. By the same token, many say that charging tax on water is unjustified at a time when Tapay's water supply is shrinking and that it makes no sense to pay the tariff because they provide the labor to maintain the water canals and reservoirs.

But even though many villagers oppose the water tax and water tariff, Tapay's water management and the maintenance of its irrigation infrastructure are increasingly becoming monetized. Many water users are migrants who do not have the time to fulfill the service as water allocator and therefore hire villagers to replace them, as Godofredo did in 2011. And as it has become more common for the water allocators to pay others to do their job, a growing number of water users now offer them cash in return for irrigating their fields (Paerregaard 2019a). The use of money is also prevalent in other activities such as *faenas* (Spanish: collective work), which are mandatory calls for community work including *yaqar aspiy*. Rather than taking a day of the calendar or spending several days traveling back and forth between Tapay and the city, many villagers and migrants prefer to pay someone else to do the *faena*.

Just as Tapay eventually embraced the organizational structure introduced by the 1969 water law, it has adopted the institutional framework dictated by the 2009 law. At the same time, it is monetizing the *regidor* office and the *faenas*. Both changes are signs that the autonomous and community-based management and maintenance of the district's water supply and irrigation infrastructure are yielding in favor of the governance model introduced by the state. Even so, resistance against the water tax and tariff continues to be strong; equally, many villagers (and migrants) still believe that it is the mountain and not the state that controls the water flow. As I discuss in the next section, the offering rituals and the notion of existential metabolism that drives them constitute a crucial element in this cosmology.

EXISTENTIAL METABOLISM

The counterpoint of the state's attempt to tax the use of Tapays' water is the tributes the villagers pay to their water sources. During my thirty-five years in Tapay I have attended such offering rituals to springs, offtakes, canals, and reservoirs on numerous occasions. Similarly, I have participated in offerings to *huakas*—that is, caves concealing ancestors' bones and sometimes also ceramics and textiles, which are thought to be the hiding places of the *gentiles* whom the villagers believe control the water flow in Tapay's springs. In fact, paying respect to Tapay's more-than-human beings is an activity integrated into the villagers' everyday lives, and as an anthropologist doing fieldwork in the district, knowing how to make *tink'a* by sprinkling or spilling the corn beer or spirits one is offered is critical to not only one's work but also efforts to not get (too) drunk. Thus, knowing the names of Tapay's most powerful mountains and other more-than-human agents enabled me to salute them one by one whenever someone offered me a drink and in this way to reduce the amount of alcohol I consumed. Of course, exaggerating the libation may catch people's attention and cause their indignation. In 1986 an old woman yelled, "He is trashing our love" when I sneaked out to pour a cup full of alcohol on the ground during a social event. But once I learned the trick of dropping the proper amounts of liquid while calling the nonhuman agents by their right names and in the right order, I managed to not only stay on my two feet for the rest of the day but also gain the villagers' recognition.

The offering at Seprigina's summit with Godofredo was therefore not a novelty to me. Still, making it to the top of Tapay's highest and most respected mountain on November 1 was a thrill because the ritual was the only event that I hadn't attended in the sequence of offerings the villagers make to Tapay's water sources during the year. People had always told me that the offering to Seprigina is essential to assure Tapay's water supply. But so are other rituals. Hence, I was curious to witness in situ how Godofredo's tribute to the mountain contributed to the idea that Tapay's multiple offerings at different sites and on different dates are linked in one coherent worldview.

In 1986 I observed that every offtake, spring, canal, and reservoir is protected by more-than-human beings that demand *pagos* sometimes several times a year.[17] People also told me that the harvest hinges on these ceremonies, which aim to appease the nonhuman agents and ensure a continuous flow of water. During my fieldwork I observed that while each hamlet made offerings to the water at different places and times, the acts were linked in a cosmological order that supposedly transcended and united what was in other ways a dispersed and fragmented system of water distribution, an order that became evident in the repetition and overlap of *pagos*. Thus, some of the *pagos* ceremonies were repeated up to several times, and in some cases the second and third offerings coincided in time and space. In the following list a complete calendar of cleaning ceremonies and offerings is presented with reference to the district's springs and offtakes:

- June 24, San Juan: cleaning of and first offering to springs in Chuccho, Pallajua, and Huilcasco, Hanansaya
- July 29, Santa Maria Magdalena: cleaning of and first offering to offtakes in central hamlet, Hanansaya

- August 5, San Clemento: cleaning of and first offering to springs in Puquio, Hanansaya
- August 8, San Clemento: cleaning of and first offering to springs in Urunja, Hanansaya
- August 24, Santa Marta: cleaning of and first offering to springs and offtakes in Cosñirhua and Malata, Urinsaya
- August 30, Santa Rosa de Lima: cleaning of and first offering to offtakes in Fure
- September 8, La Virgen de Natividad: second offering to springs in Puquio and Chuccho, Pallajua, Urunja, and Huilcasco, Hanansaya
- September 10, La Virgen de Dolores: cleaning of and first offering to offtakes in Llatica
- September 20, San Francisco: cleaning of and first offering to offtakes in Tocallo
- October 4, San Francisco: second offering to offtakes in the central hamlet
- November 1, Todos Santos: second offering to springs in Cosñihua and Malata, Urinsaya, and third offering to the mountain peak supplying the water to the offtake in the central hamlet, the two hamlets in Urinsaya, and all springs in Hanansaya

In the calendar several offerings in Tapay's moieties coincide when repeated either for the second or third time (see chart 1). Time and place vary for all first offerings, while the second offerings partially coincide in time even though they took place at the same locations as the first. As for the third, place and time merge in a single event: November 1 on Seprigina is considered the most important *paqo* of the year, when the water users act as one community, paying respect to a single power believed to control Tapay's many water sources.[18] The chart demonstrates this interplay of dates.

In the chart we observe that while the villagers make three offerings a year to the water sources in Hanansaya, in Urinsaya they only make two. The dates may seem to interrelate, in that several offerings in Hanansaya (Puquio, Chuccho, Pallajua, and Huilcasco) are repeated the second time on the same date —although at different places—and that the second offering of Urinsaya coincides in time as well as place with the third offering of Hanansaya. This falls on Todos Santos (Spanish: All Saints Day) and is considered the most important of all rituals associated with irrigation. During

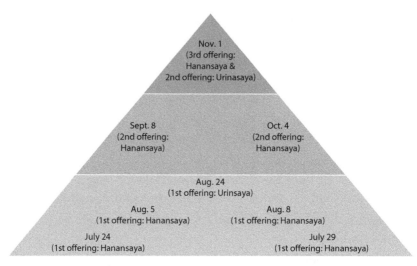

CHART 1. Tapay's offering calendar, in the Colca Valley, Peru. Chart created by author.

my stays in Tapay in the mid-1980s and early 1990s the *regidor*, together with a *paqu*, made a personal visit to the summit every year and delivered the offering directly to the nonhuman powers controlling the flow of water (see figures 3, 4, and 5). As to the four hamlets not included in the dual division, all except one made a first offering at places and times different from the rest of the district. In 1986 I obtained no knowledge of either a second and a third offering or the existence of any *pagos* in these hamlets. Hence, Paclla (which has no official offerings), Llatica, Fure, and Tocallo are not included in the model.

The model demonstrates how all offerings advance toward the same geographical spot, Seprigina, culminating at the same point in time: the November 1. The merging of time and space is made possible because the villagers on this particular day act as a united community of water users by paying respect to a single power that they believe controls Tapay's many scattered water sources. But the model not only shows that the water users believe their water sources are controlled by the same power and that Tapay's springs, rivers, offtakes, canals, and reservoirs constitute one cohesive water body. It also suggests that they imagine the water flow as a hydraulic circle that constantly renews Tapay's water supply, a notion that resonates with a more general Andean perception of the world as floating on a great sea that connects the fresh water of rivers, springs, and lakes with the saltwater of the oceans (Bastien 1985; Urton 1981). However, unlike the hydrological cycle that modern

FIGURE 3. A shaman preparing an offering to Mt. Seprigina, Tapay's highest mountain, in the Colca Valley, Peru, 2011. Photo by author.

FIGURE 4. The regidor making an offering to Mt. Seprigina at the summit, in the Colca Valley, Peru, 2011. Photo by author.

FIGURE 5. Women serving corn beer at an offering ceremony in Tapay, in the Colca Valley, Peru, 2011. Photo by author.

science claims constitutes a self-contained water flow, the Andean hydrological cycle is driven by a metabolic process that is controlled by the mountains and other nonhuman agents and that humans mediate by making offerings. But how exactly do humans do this?

In 2011 Godofredo's *paqu* confirmed what other *paqus* had told me during my fieldwork in the 1980s and 1990s: the *pago* consists of an *iranta*, of which several elements are the same as I described in chapter 1: seeds and leaves of the coca plant, *konuqa* (an herb from the *puna*), *qorilibro* and *qolqelibro* (small objects of gold and silver), an alpaca fetus and *pichuhuira* (fat from the breast of a llama), corn of three different colors, *qochayuyo* (seaweed), and starfish. By the same token, Godofredo's offering followed the same procedure that I previously had observed: some of the objects (corn, plants, seeds, leaves, fat, fetus, fish, weed, and metal pieces) are burned, others (cigarettes, coca leaves, and spirits) are chewed, drunk, sprinkled on, or dropped to the ground, while yet others (seawater, *chicha* of three different colors, and wine) are left at the offering site to be absorbed by the earth.

As a collection of the key elements of Tapay's material environment (on as well as under the ground) and ceremonial/spiritual universe (Andean as well as Catholic), the *iranta* represents the world that the villagers inhabit and that encircles their water supply. Three of the offering items are particularly conspicuous. Seaweed, starfish, and seawater all originate in the sea, which indicates a belief in deeper ties between the apparently independent springs and echoes the pan-Andean idea of Earth as floating on a sea that units the world. Many villagers affirmed the vision that the water is controlled by non-human powers, and when I asked why they made offerings, they answered because water is a living being that gives life only if it receives something in return, suggesting that Tapay's water sources constitute a cohesive system.

Although the villagers seemed to disagree when I asked them to explain the merging of their offering ceremonies, I interpret these differences as variations on a common theme rather than as competing understandings. Thus, my material suggests that people conceive of nonhuman rather than human powers as controlling water and that this belief is intricately linked to the nature of Tapay's irrigation system, fed by local springs and offtakes rather than by a single river or stream and organized by local groups of water users rather than a central water authority. Power to control water is therefore imagined as a relationship between humans and the nonhuman world personified as the mountains and the ancestors.

CLIMATE BLUES

On my return visits to Tapay in the 2010s I learned that the villagers continued to pay tribute at the district's many springs, reservoirs, and, of course, Seprigina.[19] Similarly, I found that the offering model discussed earlier was still valid and that the offering ritual had remained unchanged since 1986. However, I noticed one important change: climate change and especially Tapay's water supply had become a source of great concern. A key question in my restudy was therefore how climatic and environmental change affects Tapay's irrigation and offering practices.

The villagers have used the term *la clima* (rather than *el clima*, which is grammatically correct in Spanish) to describe the weather since I first went to Tapay. But when I did my restudy in 2011, *cambio climatico* (Spanish: climate change) had been added to the villagers' glossary. And even though few had detailed knowledge about global climate change, many complained that Seprigina and the nearby mountains (Surihuiri, Shulloqa, Bomboya, and others) were only snow covered during the rainy season and not the rest of year as they used to be. The villagers' claim resonated with my own observations and was supported by the household survey on climate perceptions I conducted in 2016, which showed a broad consensus in Tapay that climate change represents a serious threat to the district. The most notable environmental changes reported by the interviewees were rising temperatures, new insects, worms appearing in crops and fruits, irregular precipitation, and shrinking water supplies, which in different ways damage Tapay's two principal economic activities: agriculture and fruit growing.[20]

The fifteen villagers I interviewed in the household survey gave me a variety of reasons why the environment was changing, referring to anything from geological and meteorological to moral and teleological causes (Paerregaard 2013a, 2020a). While several of them adeptly employed the glossary of global warming to account for climate change, they disagreed when I suggested that it was a problem caused by the industrialized countries. Some of those interviewed argued that rising temperatures are due to the introduction of modern lifestyles in Tapay and Peru's contaminating mining industry. Others suggested that climate change is a cyclical phenomenon that people in the Andes have known for centuries and that is related to natural disasters such as earthquakes, hunger, and plagues. A man who runs a small hostel for backpackers and who takes great interest in conversing with his visitors said: "In Europe you're worried about climate change, but we've known it for a long time. Today water is scarce, but things will change again." On a similar note, a woman said that even though rising temperatures are inevitable, they are only temporary and, like other natural disasters such as earthquakes, they will be followed by better times. Another woman told me that the villagers had experienced worse changes. She said: "My grandmother recalled a year of draught when the Colca River dried up and there was nothing to eat." Another villager pointed out to me that climate change is a phenomenon that occurs locally, not globally, and that global warming is caused by the encounter between warm water coming up from the earth and the cold water falling as rain from the sky. Yet others see climate change as a sign that the world is growing old and morally corrupt. According to one villager, rising temperatures and lack of rain are omens of an apocalypse that will end the world. A similar interpretation was offered by a man who asserted that the sun no longer is protected by "the shirt" it used to wear, and because it is now "naked" it burns more. Expressing the same fatalistic worldview as other villagers I interviewed, the man said: "It doesn't even help to make offerings anymore. There is nothing we can do."

In fact, the offerings have been a contested issue for many years. During my field research in the early 1990s, a small group of Protestants called *hermanos* campaigned against not only the Catholic church but also Andean customs and ceremonial practices such as the offerings to the mountains and the ancestors, and they repeatedly tried to obstruct the villagers' ritualization of the annual cleaning of the water reservoirs and canals in Tapay (Paerregaard 1994a, 1994b). Even so, the vast majority continued to participate in these events, and in the late 1990s many of the *hermanos* gave up their struggle and reconverted to Catholicism. In 2011 I noticed that climate change, and particularly Tapay's shrinking water supply, had prompted some Catholics to question Tapay's offerings too. Nonetheless, unlike the man cited here, they continued to participate in the offerings not because they believed the tributes would produce more water but because the rituals are a custom and, even more important, because omitting to make them could bring misfortune on the *regidor* and his family. The following year when I did my restudy, several villagers reported that more rain had fallen in 2011 than in the past fifteen years. But when I pointed this out to the incoming *regidor* who was responsible for the 2012 offering, he told me that he intended to carry it out as planned, despite the climate's unpredictability and Seprigina's dubious response to the villagers' tributes. "Not because I believe it'll make any difference but because it's a custom," he said. Several villagers approved of his decision to continue the offerings. Among them were a man who pointed out to me: "We do it not because there'll be more water but to elude the anger of the mountain and the ancestors." He then reminded me of what had happened to the wife of the *regidor* who neglected to make the offering to Seprigina in 2010: "She fell ill and died. The mountain ate her."

But climate change is not the only change in town. It happens at a time when Tapay is undergoing a major transformation. Since the early 2000s the Colca Valley has experienced a tourist boom, and today the region is Peru's second most popular tourist site (after Cusco). One of the most popular tourist sites is Tapay, which has become a haven for trekkers and backpackers who see the district as an exotic relic from premodern times. During the high season from June to August several hundred tourists walk into Tapay daily, and to serve the visitors more than a dozen villagers have put up hostels and restaurants, which provide them with extra income. Most of the tourists come from Europe and other Western countries, and through the stories they tell about modernity and globalization the villagers have appropriated terms such as CO_2 omissions, contamination, and global warming and have been introduced to the modern notion of climate change. Even though many villagers disapprove of these ideas, as I discussed earlier, the global vocabulary on climate change introduced by the tourists has provided them with a new discursive framework to review and articulate the environmental changes they are experiencing. And the tourists are not alone in instituting a new climate awareness in Tapay.

When I did fieldwork in 1986 the Peruvian state seemed utterly remote. The only visible proof of its existence was the district's schools and the ID papers some of the villagers had managed to obtain. Although they celebrated Peru's day of independence (July 28) by hoisting the Peruvian flag and watching schoolchildren march while citing lyric homages to *la patria* (Spanish: the fatherland), few felt any personal connection to Peru or the Peruvian state. Arguably, Tapay's almost autonomous district governance exemplified the vernacular statecraft I referred to previously. This has changed dramatically in the past two decades. Today multiple government institutions and programs are operating in the district, offering its water committees cement, tubes, and other material to improve their offtakes, canals, and reservoirs, which not only smooths the water flow but also eases the task of cleaning the waterways. Using similar provisions from the state, Tapay's newly formed drinking water committees have installed running water in all the district's households. Most of them also have sewer access. The state has also installed electricity in Tapay, which now has light around the clock both in private homes and on streets and squares. Other servicers include a health post and a host of social programs to support people below the poverty line, the elderly, single parents, and children.[21] Perhaps the most powerful proof of the state's presence is the newly opened road to Tapay, which the regional government has financed. Though bumpy and only wide enough for one vehicle, it allows the villagers to take the bus instead of walking up the canyon to neighboring Cabanaconde to do their shopping, or to travel to other places. At the same time private agencies provide such services as cable TV and have put up a telephone antenna that enables the population to communicate with relatives in other parts of Peru and receive the daily news. Finally, Tapay's own migrants living in Lima and Arequipa put their stamp on the villagers' life when they sponsor the Candelaria fiesta in February, investing their hard-earned savings in music, food, and beer (Paerregaard 1997b, 1998). Truly the world is coming to Tapay in many ways. And there is more to come!

In December 2016 a gold mine started to operate in Tapay that has challenged the villagers' notions of water and power and altered their vision of the future.[22] Initially the mine caused much concern among the villagers, who feared it would contaminate or even exhaust their freshwater supply. In response, the villagers formed an association named Frente Defensa de los Intereses de Tapay y Anexos, which called on its migrants in Arequipa and Lima to mobilize support for negotiations with the mining company and organized two marches to the mine, situated above Tapay's main settlement at 4,800 meters not far from Seprigina. To back the villagers' protests and affirm their concern for Tapay's environment, migrants arranged similar events in the city of Arequipa. The company, on the other hand, has tried to gain Tapay's support by offering employment to almost two hundred villagers and migrants and sponsoring several public works in the district, including the construction of new school buildings and new municipal offices, installing water pipes from some of Tapay's springs to enhance its water discharge, the purchase of a bulldozer to repair Tapay's new road as well as a bus (and a paid driver) to transport children from the district's smaller settlements to school. Finally, and most importantly, it has agreed to construct a channel to direct water to Tapay from a nearby river (Molloco), a project that is in progress and that promises to enhance the district's water discharge significantly.[23] The mine's investments in Tapay's infrastructure and, not least, the employment of its villagers and migrants as workers have changed the image that particularly young villagers have of its interests and activities. Its presence, however, has also created deep divides within Tapay's population, and although the prospect of a new channel generates hope for the future among the young, many senior villagers still fear the mine will contaminate its water.

ADAPTATION CAVEATS

My interviews with Florencio, Mauricio, and Godofredo capture the confusion and the division that climate change creates among the villagers and illuminates the way it intersects with other changes such as the gold mine's presence in the district. Having known the three men since I did my first fieldwork in Tapay in 1986, I have observed how their belief in mountains and ideas of offerings shape their climate perception and how the mine's impact on the water supply challenges their ideas of climate change adaptation.

In the late 1980s Florencio came under the influence of Tapay's small but strong evangelist community, and following his father's example he converted to Protestantism. Some years later Florencio was elected mayor of Tapay, an office associated with not only personal prestige but also expectations to participate in the villagers' many drinking parties, which challenged his Protestant faith and struggle to combat alcohol. To manage this dilemma Florencio relapsed into Catholicism one year into his three-year term as mayor, only to reconvert to Protestantism after he had completed it. When Florencio was appointed *regidor* in Tapay's main hamlet in 2009 he once again faced the difficulties of occupying a public office as an evangelical in a Catholic Andean community, but this time he decided to stay loyal to his faith and refused to serve and drink alcohol at the *yaqar aspiy* fiesta, where his wife, who also is a converted Protestant, only offered the participants food. And more important, he refrained from paying tribute to Seprigina. Florencio's neglect of Tapay's ritual traditions caused a disturbance among Tapay's Catholic majority, who worried that his defiance of Seprigina would provoke its anger. But during an interview in 2011 Florencio told me these worries were groundless. He said, "I don't believe in these traditions, and I don't fear the mountain. And by the way, there's less water now than before even though other *regidores* have made offerings." Still, Florencio shared the concern that the climate is changing and that the growing water scarcity jeopardizes Tapay's subsistence. In 2016 Florencio said to me that the newly established gold mine could make the situation worse. Yet rather than opposing it Florencio had taken a job in the mine. He argued: "The mine is here, so why not get the best out of it? They pay me well and give me good working hours." As to environmental change, it was Florencio's opinion that the villagers must adapt to the new situation and find new ways of making a living. He said: "There are no children or young people anymore, only families with two persons and we don't have many needs." But while Florencio continued to think that offerings can do little to undo climate change, he had changed his view of the mine on my last visit to Tapay in 2019. Florencio told me he didn't work in the mine anymore and that he no longer approved of its activities, which he thinks contaminates the district's water supply. The mine is "like an octopus with long arms that'll eat you" and "it eats everything and leaves the mountain empty," Florencio explained me.

Mauricio has lived most of his life in Tapay except for the seven years he spent with his wife in Lima when they were young and their children were small. Today they live in Tapay, making an income by cultivating the land and providing lodging for tourists in their small hostel. In 2010 Mauricio took over from Florencio, serving as *regidor* in Tapay's main hamlet, and in 2011 he served as head of Cosñirhua's drinking water committee, which takes care of the settlement's drinking water infrastructure and charges the monthly fee the villagers pay for their water consumption. Mauricio's community work has given him a firsthand insight into Tapay's water crisis, which he finds has gone from bad to worse. In 2011 he told me: "Water has always been scarce but in recent years there isn't enough to plant all our fields, and many have abandoned them all together and gone to other places." Drawing on his own experience, Mauricio explained: "The *regidor* fills the reservoirs at night and allocate water to the fields during the day as always but now the reservoirs are not full anymore in the morning so there isn't enough water for everybody." Yet when asked whether Tapay's growing water scarcity is related to climate change, Mauricio claimed that the villagers have known it for a long time. He said: "Things are difficult now, but they'll change again. My grandmother told me that when she was young there was a drought that lasted several years. It reached a point when the Colca River dried up. The only water available came from the springs." Mauricio also recalled that Tapay suffered from a terrible plague when he was a child and that the mice ate the harvest that year. He added: "The climate is cyclical. After some years with drought, plagues and other misfortunes, there'll be rain and good harvest again." Meanwhile, the villagers must continue to make offerings to Seprigina, which Mauricio said still has the power to punish them. When I interviewed Mauricio again in 2019, he recognized that Tapay's water scarcity had worsened but also claimed it was unrelated to environmental change in the district and climate change in other parts of the world. Interestingly, like Florencio, Mauricio had altered his view of the mine since it began to operate in late 2016. But while Florencio first embraced it and later decried it, Mauricio changed from disapproving of to welcoming it. In 2013 when the mining company made its first appearance, Mauricio fiercely opposed the mine, and in the following years he played a leading role in organizing the protest marches against it. But when I interviewed him in 2017, he told me that he had changed his mind. He said: "My son is now working in the mine making good money. If I had the time, I would also consider taking work there." Mauricio's view of climate change and offerings, on the other hand, remained unchanged. He reiterated that global warming is inevitable and that Tapay must bring gifts to Seprigina to enhance its water supply.

Godofredo spent his childhood in Tapay before migrating at the age of fourteen. Like many other young villagers, he went to Peru's cities to study and work. After marrying a woman from Cusco, he settled in Arequipa,

Peru's second largest city. To provide for the family, he worked in different parts of Peru including the jungle, where he lived for several years. From time to time Godofredo visited Tapay to look after his mother and take care of the family's properties. When he retired in 2004, he and his wife decided to return to Tapay, where they have constructed a small tourist hostel with the help of their children. To supplement his pension and the income the hostel yields, Godofredo cultivates the land he has inherited and keeps a few animals. When I met him in early 2011, he said: "We live a nice and quiet life here and we have all we need." The year before, Godofredo had been appointed *regidor*, but rather than carry out the task himself as Florencio and Mauricio did, he paid someone to do it, which saved him time to prepare the offering ritual to Seprigina and organize the *yaqar aspiy* fiesta. "Allocating water only gives you problems while the ritual and the fiesta yield respect and prestige," Godofredo told me. He explained: "More and more migrants come back to visit Tapay during these activities. *Yaqar aspiy* has become a moment to reunite with the family and forget the trouble people go through in the city." The growing interest in *yaqar aspiy*, however, has made the event a reason for concern as providing food and beer for the many participants can be a costly affair. While Godofredo thinks it is important to make offerings to Seprigina, he worries little about climate change and Tapay's water shortage. When we met again in 2017 Godofredo said: "Everybody talks about climate change but it's not a big problem here. There is less water now than before but there are also less people. Most families with children are in Lima and other places and only old people live here now." Thus in Godofredo's eyes climate change represents no real threat to Tapay. He pointed out to me: "Only a small group of villagers plant their fields today. Most have other incomes. Off course, we need drinking water for our hostel but that is not a problem." He substantiates his climate skepticism by referring to other changes that currently are transforming Tapay and impacting its environment. Godofredo said: "Everybody talk about climate change but there are so many things going on these days. Now there's a goldmine in Tapay. I really don't know whether it is the climate or the mine that cause water scarcity."

DREAMING OF WATER

On my latest visit to Tapay in 2019 people told me that the *regidores* were still making offerings to the district's many springs, offtakes, and reservoirs as well as to Seprigina. However, other very visible changes were going on in Tapay. The gold mine had put up a compound housing more than two hundred workers who were working on the construction of the promised channel. Although the workers lived apart from the local population, the mine's engineers and technicians were based in Cosñihuar, Tapay's largest hamlet, where they stayed with villagers, who made a living offering them accommodation and food. During the day a flow of trucks and four-wheeled SUVs rushed in and out of the hamlet, which had only become familiar with the presence of cars in Tapay after the road from Cabanaconde was completed in 2015. The presence of the many engineers and workers sent a tangible signal to the villagers that the mining company was keeping its promise, which had a notable effect on their perception of it. Comparing the talks that I had with some of the villagers about the mine with the conversations we had had both before and after it opened in December 2016, I found that several had changed views. Some who previously had been critical of the mine were now positive, while others who initially embraced it had become skeptical in 2019. Clearly Florencio and Mauricio were not alone in altering their perceptions of the mine, whether from positive to negative or the other way around, and neither was Godofredo the only villager to confuse the impacts the mine and the climate were having on Tapay's water supply.

Still, it was my sense that the channel's prospect of enhancing Tapay's water supply had opened a new window of imaginations and expectations, which are critical drivers of the effort to adapt to climate change. Unlike conventional development that makes changes for the better, climate adaptation implies changes that people didn't ask for but that nevertheless are inevitable. To engage marginal populations in adaptation initiatives it is therefore of paramount importance that they feel there is hope. In Tapay the readiness to adapt to climate change is inextricably linked to the dream of water and the new opportunities and livelihoods it can generate, a dream that has existed for a long time. Since I did my restudy in 2011, Peru's Ministry of Agriculture has organized campaigns in Tapay to instruct the villagers how to protect fruits against bugs, and although climate change hasn't been part of the campaigners' glossary, they have contributed to the creation of a new climate awareness and, more important, hope in the future. On numerous occasions villagers have told me they plan to benefit from the warmer climate to change fruit growing by replacing their old apple, pear, peach, and fig trees with mandarin, apricot, and orange trees. Of particular interest, they have pointed to the prospect of planting avocado trees in the *botaderos*, the dry slopes above Tapay's settlements that traditionally have been used as grazing grounds for animals during the rainy season and more lately for cochineal production. The villagers have explained to me that if irrigated regularly, these otherwise barren areas could be suitable for producing avocados on a large scale, which would lead to a major shift in Tapay's economy from agriculture directed toward self-subsistence and fruit used as an exchange item in the regional bartering market to commercial production of primarily avocados.[24] All they need, the villagers have told me over and over, is water.

Paradoxically, climate change impacts Tapay at a time when not only is the district's population both shrinking and aging, but the few villagers who are left have been connected to the outside world by road and may look to a brighter future with more water and the possibilities of producing cash fruits for the national and world market. Regarding this book's topic, this raises the question of whether the opening of the new channel and the change of Tapay's water supply it involves will alter the villagers' perception of water metabolism and the power that controls the water flow. In other words, will Tapay continue to make offerings to Seprigina, or will it stop performing the ritual? Alternatively, will the villagers pay tribute to the channel instead of the mountain and in this way recognize the mining company as a new agent in the metabolism that moves water? It's still premature to answer these questions, but the next chapter may give a hint of what could happen in Tapay in the coming years.

THREE

Cabanaconde

THE HOLE IN THE CHANNEL

"They were eleven men, and it took three days and nights before the water came out," Nicolás said, describing to me how in March 1983 a group of villagers made a hole in the Majes channel that crosses the territory of Cabanaconde. I ran into Nicolás when I visited a museum that an organization called Auto Colca had built in Cabanaconde to honor a mummy found in the district in 1995.[1] Nicolás was the guardian of the museum, and as this was empty (the mummy is kept in Arequipa; discussed later) and there were no other visitors, he had plenty of time to talk. He continued: "The intellectual leader of the action was a schoolteacher from Cabanaconde but he didn't take part of it, neither did he go to jail afterwards." As the schoolteacher, Nicolás was not one of the eleven men, but as municipal secretary in 1983 he followed the event closely and became personally involved in its aftermath. He admitted to me that he initially objected to that schoolteacher's idea but that he changed his mind after Cabanaconde's mayor and his assistants had approved of it. Nicolás recounted: "They worked when it was dark, and it took two days to make a small hole with the tools they had brought. But the hole was too small so on the third night they returned with dynamite and that worked. The water flowed right into the Hualca Hualca River."

The reaction from the regional authorities was prompt. "The same day *los republicanos* [a special police unit in Peru] arrived and arrested the mayor, the governor and two municipal officials including me and took us to Chivay [the regional capital]," Nicolás told me. However, the men's arrest only lasted two days. As a member of the political party in government (Acción Popular), one of the eleven men who made the hole in the channel went to Arequipa to ask the governor, the region's highest political authority, to release the four detained villagers, which he then did. Upon their return to Cabanaconde the men were received as heroes, and together with the eleven activists they are remembered as an example of community unity and resistance.

The action in 1983 was a turning point in Cabanaconde's contemporary history. It took place in a year when Peru was hit badly by the El Niño current, which caused flooding along the northern coast and severe drought in the southern highlands. Elderly villagers recall that the harvest failed because it didn't rain and because the meltwater from Hualca Hualca, their principal water source, almost dried up. Because of chronic lack of water they had unsuccessfully tried to reconstruct an old canal from a local watershed called Huátac to enhance Cabanaconde's water discharge for several years. The same year the Peruvian state completed the construction of the Majes channel, one of Peru's largest and most costly water infrastructure projects, which transports water from the upper Colca River to the nearby coastal desert (Paerregaard, Ullberg, and Brandshaug 2020; Stensrud 2021). The project's aim was to alleviate the pressure on land in the Colca Valley and neighboring areas by offering thousands of highland peasants small lots of irrigated land on the plains of Majes.[2] On its way to the coast the channel passes through the territories of the districts on the left bank of the Colca River, including Cabanaconde, which eventually took action, claiming its right to the channel's water. Soon after the incident the channel's administrators decided to allow Cabanaconde and the rest of the districts on the left bank access to the water. The rest is history, which I account for later.

In the years that followed Cabanaconde's epic confrontation with the state, the community altered not only its water management but also its ritual practice. By shifting from existential metabolism, as practiced in Tapay, to contractual metabolism, Cabanaconde has made a contract with the state that offers the community a safe and steady water supply in return for paying water taxes and tariffs. On a voluntary basis, the presidents of Cabanaconde's water user committees also make offerings to the five valves in the Majes channel that direct water into the community's irrigation infrastructure. Unlike existential metabolism, which seeks to smooth a water flow that is controlled by nonhuman forces, contractual metabolism is an arrangement between human stakeholders who hold different degrees of power to negotiate and define its terms. Before discussing how contractual metabolism works, however, an introduction to Cabanaconde's demographic history, population movements, and water governance is pertinent.

Situated across from Tapay in the bottom of the Colca Valley (see figure 6), Cabanaconde is the second largest district in the region, with more agricultural land than other settlements. The bulk of the villagers make a living by cultivating the land between 2,200 and 3,600 meters, but a few also grow fruit on the banks of the Colca River at 2,000 meters or keep animals on the *puna* above 4,000 meters. Cabanaconde's population is concentrated in one big settlement that was connected by road to the rest of Peru in 1965. In addition to Cabanaconde's confrontation with the state in 1983, several other developments have changed life in the district. In recent years the district has become a center of attraction to a growing number of tourists, who use it as a stopover when visiting the canyon and neighboring Tapay and provide the villagers with a new source of income as hotel and restaurant owners and tourist guides. Simultaneously, the introduction of parabolic antennas and other modern media practices and communication technologies, combined with electricity service around the clock, has allowed villagers to watch the news both nationally and internationally. Another important change was the installation of a permanent telephone service and more recently also internet service, which has generated new urban-inspired consumption practices that are rapidly transforming Cabanaconde's traditional rural lifestyle (Gelles 2005).

Cabanaconde takes pride in its premodern history. During the Incan era the Colca population was divided into two polities: Collaguas and Cavanas. While Tapay belonged to the former, Cabanaconde constituted the center of the latter. Both polities lost importance after the Spanish conquest, and during the early colonial period two *encomiendas* (Spanish: estates) were established in what formerly was Cabanaconde Hanansaya and Cabanaconde Urinsaya. In the late sixteenth century when the colonial state gathered Peru's native population in *reducciones*, Cabanaconde was established as the nucleated settlement we know today (Gelles 2000, 29–34). Outsiders have been present in Cabanaconde since the Spanish conquest, mostly in the form of Spaniards and *criollos*, who settled to exploit the land and the nearby mines of Caylloma.[3] Their presence formed the basis of the ethnic identities between insiders (*indios*) and outsiders (*mistis*) that continued to divide Cabanaconde's population long after Peru's independence in 1821 and that only began to crumble in the second half of the twentieth century. Gelles writes that these outsiders "created the conditions for the survival of a small

FIGURE 6. The district of Cabanaconde, in the Colca Valley, Peru, 2011. Photo by author.

enclave of literate, powerful criollo families—called 'españoles' (Spanish), 'blancos' (white), and 'mistis' locally—who ruled Cabanaconde and continued to intermarry" (Gelles 2000, 123). The influence of these families was reinforced after the municipal district was established in 1857, which enabled their members to fill the offices of *alcalde* and *comisario* (Spanish: commissioner) and in this way oversee Cabanaconde's irrigation management and other important public works undertaken by the native population. But as in Tapay, the ethnic divide began to blur as the outsiders assimilated and over time became insiders, and it was only by keeping ties to the larger regional elite that a few families managed to claim a separate ethnic identity as Spanish descendants and dominate Cabanaconde's political life until the 1960s (Gelles 2000, 124). The official recognition of Cabanaconde as a *comunidad campesina* (Spanish: peasant community) in 1979 was the final nail in the coffin of the dominance of these families.[4] However, even though ethnicity has lost importance as a symbol of social rank, new forms of social differentiation have emerged. In terms of economic income Cabanaconde is still a relatively homogeneous population, but one activity divides the people: migration.

For more than half a century Cabanaconde has experienced continuous out-migration. Initially this migration was directed toward Arequipa and Lima, Peru's two major urban centers, and by the 1940s Cabanaconde had established migrant associations in the two cities. In the second half of the twentieth century migration grew steadily, and in 1987 the migrant populations of Arequipa and Lima were approximately one thousand and three thousand respectively (Gelles 2005). The demographic impact of Cabanaconde's out-migration can be read in the development of the district's population, which grew steadily throughout the twentieth century until it peaked forty years ago.[5] The 2007 census counted 2,842 and the 2017 census only 1,865. The main cause of this decline is out-migration, which has had little impact on the gender distribution of the population but is notable in age distribution.[6] Although not as prominent as in Tapay, the tendency is the same: Cabanaconde is rapidly aging.

An important contributor to this development has been the migration flow to the United States. Spearheaded by a handful of young women, Cabanaconde initiated a transnational network in the early 1970s that gained momentum in the 1980s and that over the past three to four decades has pulled hundreds of villagers (women as well as men) toward the same end station: Washington, D.C. By 2005 the migrant community in that city had approximately five hundred members. Today it is close to one thousand (Paerregaard 2010a, 2017). But Cabanaconde's migrant colony in Washington is not only large; it is also well organized and has carved out its own niche within the city's Peruvian and Hispanic community (Paerregaard 2005, 2014a). In 1983 the migrant colony formed the Cabanaconde City Association (CCA), which arranges annual soccer tournaments and social events to collect funding for projects in Cabanaconde such as the construction of a bull-fighting stadium and a free lunch service for the elderly. Many migrants also remit money to their relatives in Cabanaconde and other places in Peru (Paerregaard 2015a). Moreover, every year community authorities, schoolteachers, parents' organizations, and other associations in Cabanaconde request help from the CCA to provide them with equipment (instruments, computers, and uniforms) and furniture (chairs, benches, and tables), to repair the walls and roofs of public buildings, and to improve irrigation canals.[7] Individual community members also ask the CCA for help to pay hospital and medical bills when family members fall ill or suffer from accidents. However, mutual distrust between CCA and the district authorities has crippled their cooperation, and while many migrants think it difficult to hold the mayor and his administration accountable for the funding they provide, the latter say that the CCA only supports projects it finds relevant and that it rarely listens to the villagers' wishes (Paerregaard 2010b).

Even though many migrants no longer have close relatives in Cabanaconde, they feel a need to affirm their community membership.[8] Quite a few own land in the district, which they either cultivate in partnership with other villagers or rent to the growing number of outsiders from places such as Cusco and Puno who have moved to Cabanaconde in recent decades. Some migrants also spend their savings on constructing modern-style houses in Cabanaconde, which they use on occasional visits. One family has even taken out a loan to construct a luxury hotel with a swimming pool and a casino in Cabanaconde, which they hope other migrants will use when visiting.[9] More importantly, every year migrants living abroad (but sometimes also from Lima or Arequipa) volunteer as *devotos* of Virgen del Carmen, Cabanaconde's patron saint, who is celebrated in mid-July. By sponsoring the fiesta they create an image of themselves as Cabanaconde's successful "children" and gain the respect of not only the villagers but also other migrants (Paerregaard 2017). The sponsorships are read as a sign of migrants' *fieldad* (Spanish: loyalty or devotion) to Cabanaconde and as a demonstration of their *voluntad* (Spanish: social volition) to support it (Paerregaard 2015b).

Migrants' competing sponsorships have transformed Cabanaconde's fiesta into a transnational event that entices migrants to visit their home district but that also makes many locals feel estranged. In the past ten years the cost of sponsoring the fiesta has skyrocketed and today it can easily pass US$100,000, an astronomical amount of money to most villagers.[10] In 2014 a successful businessman from Cabanaconde who lives in Chile tried to outdo all previous *devotos* by sponsoring what people say was the costliest fiesta ever. Many villagers disapproved of the man's excessive expenditure and deplored the escalation of the *devotos'* rivalry, which has developed into a competition between migrants from different countries. Some even contended that the trembling that occurred during the fiesta was an omen that the *devoto* had exaggerated the sponsorship. One villager told me: "He just wanted to show his money. I didn't feel welcome at all and many of us only joined the music when it was dark, and no one could see us."[11]

While migration has been an opportunity for young villagers to achieve social mobility inside as well as outside of Peru, it has contributed little to the villagers' wealth.[12] A survey on Cabanaconde's socioeconomic situation and social and environmental vulnerability that I conducted in 2011 is evidence of this (Paerregaard 2018a).[13] It showed that farming and herding are the principal occupation for the vast majority, and even though a small number of villagers have other sources of income, only a tiny section lives entirely off nonagricultural activities.[14] The survey also showed that three-quarters of the population were born in Cabanaconde and that five out of six are return migrants.[15] Finally, it demonstrated that household participation in the activities organized by Cabanaconde's community is extensive and that almost half of the heads of household are or have been active members of its irrigation committees. In other words, even though the population of Cabanaconde is almost four times as large as Tapay, and even though it receives support from a large transnational migrant community and has easier access to Peru's urban centers and the national market than Tapay, the two districts are facing similar challenges: continuous out-migration, an aging population, and lack of economic opportunities. Even so, Cabanaconde and Tapay differ in one important aspect: their access to water.

DISPUTING WATER

The bold move Cabanaconde made in 1983 to access the water of the Majes channel stands as a hallmark in the villagers' memory. However, the versions of how it happened are multiple. Reconstructing the event, Gelles writes: "The Majes project caused much anger in Cabanaconde to whom it was a token of the policy that the Peruvian state had pursued since colonial times, ignoring vital interests of Andean peasants such as water to irrigate their fields" (2000, 64). He continues: "In a desperate attempt to gain a right to tap water from the channel, the Cabaneños wrote a letter to President Belaunde to ask for his help. The President never replied and his silence only added fuel to their anger" (2000, 64), which prompted the eleven men to take things into their own hands. "A police contingent was sent to the community, but when they arrived the entire community confronted them," Gelles reports (2000, 64). He contends that Cabanaconde's astuteness prompted the regional authorities to eventually yield to its claim and that "out of fear of further conflicts, Autodema (Autoridad Autónoma de Majes)—the administrative unit of the Majes Canal since 1982—finally agreed to cede 150 liters per second to Cabanaconde" (2000, 65).[16] Moreover, Gelles writes, having "threatened to take similar action against the

Majes Project," the rest of the districts on the Colca River's left bank were given access to the channel too (2000, 65).

While doing field research in Cabanaconde's migrant colony in Washington, I got to know one of the eleven men who had made the hole in the channel. My conversation with Isidro, who had lived with his family in the United States for more than a decade, about the event differed little from Gelles's reconstruction, just as Nicolás's account supported it. However, when interviewing some of the engineers who worked for Autodema, the agency that administers the Majes channel, in the early 1980s, I heard other versions of what led to its allowing water to Cabanaconde. In 2011 one of them, who in 1983 had worked at ALA's office in the coastal town of Pedregal, told me that Cabanaconde already had been promised water when the villagers made the hole and that the project in fact had planned to supply all the districts on the Colca River's left bank before the channel opened. In 2016 another engineer who had been working for Autodema since in its creation in 1982 confirmed this version, pointing out that it wasn't the villagers' action in 1983 that led to the provision of water to Cabanaconde and the other Colca districts. It was Autodema's own decision, the man said. Gelles partly affirms the engineer's testimony, writing that the Majes Project had promised Cabanaconde to provide 1,000 liters per second to recover several thousand hectares (2000, 62). But, Gelles concludes, Autodema's commitment never materialized: "Promises were made but not kept" (2000, 61).

Almost forty years after Cabanaconde's action, it is difficult to tell whose version is right and whose is wrong. Even so, regardless of their validity, the competing stories of what prompted Autodema to give Cabanaconde access to the channel shed light on the frictions that shape Peru's water governance and the epistemic as well as cultural discrepancies that divide the country's water engineers and water users. The Majes Project encompasses an assemblage of channels, canals, reservoirs, technologies, and devices that link the Peruvian state and its water administrators and experts to a range of water users with different social histories and identities (see map 2). It was created to promote agriculture and economic development on the coast, but by including Cabanaconde and other Colca districts the channel connected the project's own hydraulic systems to a labyrinth of smaller irrigation infrastructures that are managed and maintained by a myriad of water user committees according to their own customs.[17] Hence, even though the channel's users are part of the same water supply chain, and even though they all tap water from the same source, the project's multilayered division perpetuates

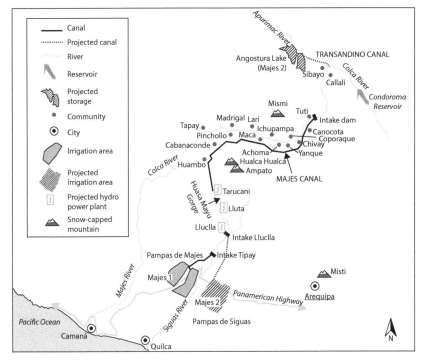

MAP 2. Map of Majes channel, in the southwestern highlands of Peru. Reproduced from "Figure 1. The Colca-Majes-Camaná catchment and the Majes Irrigation Project," in Vera Delgado and Vincent (2013), with the authors' permission (map by Juana Vera Delgado).

the users' perception of their local infrastructure as discrete and autonomous units. Moreover, it affirms the users' notion of the channel's valves that give them access to the hydraulic system as devices they have a moral right to manage. The water users' image of the project's infrastructure as a conglomerate of self-contained irrigation systems and the valves as self-regulating gateways is reinforced by their water values, which attribute importance to water as not only a material and economic asset, as the project's experts and managers do, but also a social and cultural good (Paerregaard 2019a).

The Majes infrastructure's inherent social friction comes to the surface in years of drought, which happened in early 2016 when the rain arrived late. As the water shortage forced people in the Colca Valley to continue irrigating in January and February when the crops normally are rain fed, tensions were building up between the channel's stakeholders. Following is a description of how different user groups negotiate their rights to the channel's water and how its experts and managers try to mediate their competing claims.[18]

In February 2016 I attended a so-called multisectoral committee meeting at ALA's office in Pedregal—the main town of the Majes plains—where the channel's administrators and stakeholders meet once a month to coordinate its management and in years of water stress to reach an agreement on how to reduce water consumption.[19] In years of rain each district and water committee receives a prefixed amount of water, which as in other parts of Peru is estimated as a water discharge per hectare per irrigation season. Different crops are granted a specific amount of water according to an officially sanctioned cropping pattern, which can vary for different valleys and irrigation systems (Vera Delgado and Vincent 2013). These figures are then converted into l/s/ha (liters per second per hectare) given the variations in irrigation season length (Vos 2005), but as their basis is unclear to local water users, the users often question them, claiming more water to be able to irrigate more frequently and irrigate as much land as possible (Vera Delgado and Vincent 2013).[20] During the meeting the water experts' predictions of the water shortage the drought was going to cause in 2016 prompted representatives from one of the three water user groups on the Majes plains to contend that the water committees in the Colca Valley were tapping more water from the channel than agreed. However, as the representatives of the Colca water user organization were absent at the meeting and the allegation lacked proof, the water experts and managers decided to inspect some of the valves that supply the water users in the Colca Valley.

Two weeks later I joined the inspection by three engineers from ALA and Autodema of one of the two valves that the district of Yanque uses to take water from the channel.[21] On the trip to the valve, one of the engineers recalled that this had been wide open when he inspected it the previous week. The other engineers affirmed that they had had similar experiences in other districts, pointing to the problem that the Junta de Usuario del Valle de Colca (the Colca communities' water user organization; see chapter 1) keeps one of the two keys to the channel's valves (Autodema has the other key) and that the association often let the water committees use it to regulate the water discharge without ALA's or Autodema's approval. On our arrival at the valve the engineers used a stick, a meter, and a table to measure the valve's water flow. Their estimate showed that the water flow and the valve's setting complied with the limits prescribed by Yanque's license of 150 liters per second. Nonetheless, one of the engineers noticed that the soil was wet, which he said was a sign that water had been spilled on the ground. The men took this as proof that Yanque had taken more water than it was entitled to and that the allegation against the Colca water users presented at the meeting in Pedregal was correct. At the next multisectoral committee meeting in Pedregal the engineers reported their findings, but as the Colca water users' representatives once again were absent and other urgent matters required the attention of the participants, the case was dropped. The case shows that while the Colca water users rarely attend the meetings in Pedregal, where they feel their voices are ignored, they use more shrewd methods to access the channel, profiting from their location at its upper section, which allows them to take water first. Cabanaconde's action in 1983 still stands as a model.

When the channel opened in 1983 Cabanaconde was allowed water from a valve situated at Tomanta, which is the intersection between the channel and the Hualca Hualca River that transports water from Hualca Hualca to the district's principal agricultural area, La Campiña. Today Cabanaconde taps 530 liters per second from a total of five valves, which apart from La Campiña supply 1,000 hectares of abandoned terraced fields that have been put back into production, doubling Cabanaconde's land base (Gelles 2000, 66–68).[22] According to my 2011 survey, the district's households (including members as well as nonmembers of Cabanaconde's peasant community) now have an average of 1.75 hectares of irrigated land, compared to less than 1 before Cabanaconde gained access to the Majes channel, which is more land under irrigation than any other Colca district (Gelles 2000, 59–68). While land in La Campiña is privately owned, the newly irrigated fields are communal land to which members of Cabanaconde's peasant community are granted the right of usufruct during their lifetimes, including the right to rent it to others, which many migrants do.[23] In the past two decades the state and the regional government have also financed the improvement of the district's irrigation canals and the construction of six water reservoirs that spare the water allocators the work of irrigating at night and minimize the waste of water.[24] Moreover, the Peruvian state is planning to add more volume to the Majes channel, which could enhance Cabanaconde's water discharge.[25]

Following the 1983 action, Cabanaconde's water management has undergone a major change. When Paul Gelles did field research in Cabanaconde in 1987, water users in La Campiña employed a dual division to allocate water (Gelles 2000, 57). Gelles reports that La Campiña's irrigation clusters were divided into two moieties called Hanansaya and Urinsaya, which each elected its own water allocators, called *controladores* (Spanish: controllers). Starting upstream, the two *controladores* engaged in a fierce competition, moving downstream to finish the allocation of water in their moiety first, a showdown that was repeated in the four irrigation rounds (Gelles 2000, 98–117). Older villagers recount that the competition served as a means of saving water and ensured that everybody including those having fields downstream received an equal share. Nevertheless, they also recall that people fought over the smallest drop of water. Indeed, many say that serving as the *controlador*, a duty all males had to assume once in their lives, was the most onerous office they had ever occupied in Cabanaconde (Gelles 1994, 248). Similarly, they relate that during the rainy season from January to March, when the villagers were free to take water from the canals and irrigate their fields, disputes over water were very common, particularly in years with little or no rain.

According to Gelles, up through the twentieth century the Peruvian state made numerous attempts to introduce a state model of water governance, with the aim of making water distribution more efficient (Gelles 2000, 69–74). A small group of villagers supported the state's attempt to modernize water management, but the majority has opposed it, pointing to the advantage of using a dual model that encourages competing *controladores* to make their utmost efforts to reduce water waste. Reporting from his 1987 study Gelles writes: "Although the state model has gained ground over the years, many aspects of the local model remain firmly entrenched" (2000, 71). However, when I did fieldwork in Cabanaconde in 2011 the water users in La Campiña had replaced the communal with a state model of water governance.[26] A locally elected water committee administers the new system, selling *tikets* (Spanish: tickets) to the villagers that authorize them to claim water from the *controladores*. And while the village authorities previously appointed these *controladores*, the water committee now hires them. A similar model has been introduced in the other agricultural areas that are irrigated with water from the Majes channel. Moreover, in contrast to the old dual model that was based on a principle of competition and followed Cabanaconde's traditional dual classification, in the new system the *controladores* allocate water sequentially, that is, from one plot to the next, using the *de canto* method and starting upstream. Finally, the bulk of Cabanaconde's water users pay both the water tax to ALA in Pedregal and the water tariff to the Junta de Usuarios in Chivay. Unlike Tapay, then, Cabanaconde has both recognized the state as its legitimate water supplier and adopted its model of water management, which has had repercussions for its practices and perceptions of the relation between water and power.

In 1995, twelve years after it had made a hole in the Majes channel, Cabanaconde again made headlines, this time not only nationally but also internationally, because of an event that surprised the villagers as much as the rest of the world. Searching for relics from Cabanaconde's pre-Hispanic past, Johan Reinhard, an archaeologist, and Miguel Zárate, a Peruvian mountaineer, found a frozen mummy on Mount Ampato, the tallest mountain in the region, situated next to Hualca Hualca. I wasn't present myself when the discovery was made, but Paulino, who as *gobernador* in Cabanaconde had authorized the expedition and rented Reinhard and Zárate mules to carry their gear and provisions, provided a firsthand account. In 2011 he told me that two men had returned several days earlier than planned. "Zárate came rushing into my office and said: 'trago' [booze]," Paulino recounted. He continued: "It took Zárate some time to get himself together but then he talked like a madman." However, it was only after Reinhard arrived with the mummy that Paulino realized the uniqueness of the discovery and understood the urgency of saving the mummy, which had already started to melt and decompose. "As *gobernador* it was my responsibility to act so I gave them permission to immediately take the mummy to Arequipa," Paulino told me. The mummy was later identified as the body of a young girl sacrificed by the Incas five hundred years ago. Due to its well-preserved condition the revelation of what has become known as the Inca Ice Maiden or Mummy Juanita caused a sensation in the scientific world, and it was chosen as one of the world's top ten discoveries by *Time* magazine (Gelles 2000, 80). In the following years the mummy was displayed in Japan and Washington, D.C., where President Bill Clinton declared his admiration for the girl's beauty.

The discovery of the mummy, which today is exhibited in a museum in Arequipa, placed Cabanaconde on the world map but caused concern among many villagers. As a symbol of the most precious gift humans can give the mountain deities, or *cabildos* as Ampato, Hualca Hualca, and other nearby mountains are known in Cabanaconde, Juanita reminds the villagers of their own past and what their ancestors were cable of: the sacrifice of their own children. While the cruelty of such an act causes the indignation of some, others think Juanita demonstrates the greatness of Cabanaconde's cultural history.[27] Still others distance themselves from the idea of human sacrifice but welcome Juanita as an icon that Cabanaconde can use to attract tourists and the attention of the surrounding world. The villagers also disagree about what to do with the mummy. One villager said to me that the removal of the frozen body had caused the anger of the mountain deities and brought bad luck to Cabanaconde. I also talked with villagers who contended that the two discoverers did wrong in bringing the mummy to Arequipa and blamed the district authorities for not claiming its return. Another villager interpreted the discovery quite differently. He pointed to the fact that it was the eruption of the nearby volcano of Sabancaya and the melting of Ampato's ice cap where the mummy had been buried that had made it appear and led to its appearance. In other words, it was a natural disaster and the subsequent rising temperature, not the discoverers' interference, that had brought Cabanaconde's past to the light of the day. Reinhard and Zárate had merely picked up the mummy, saving her from thawing and decomposition.

As described in the chapter's opening scene, I ran into Nicolás and collected his story about the 1983 action while visiting the museum Auto Colca has constructed in Cabanaconde to exhibit Juanita. The mummy, however, has never been returned, and few expect the museum ever to house her because of the technical difficulties of preserving the frozen body. When I asked Paulino and other former district authorities whether they have had second thoughts about the permission they gave in 1995 to bring the mummy to Arequipa, their reply was no. Twenty years after the discovery, Paulino (who sadly died due to cancer shortly after our conversation) said, "It was the right thing to do. The mummy would have melted in Cabanaconde." Even so, Juanita's legend is alive. In 2016 the mayor of Cabanaconde suggested to me that the museum should be arranged as an introduction to not only Juanita's but also the district's history. So far the mayor's vision has not materialized. Upon my most recent return to Cabanaconde in 2019 the building was still empty, reminding the visitors how Auto Colca spends the entry fee they pay to enter Cabanaconde.

The 1983 action and the discovery of Juanita in 1995 represent two very different turning points in Cabanaconde's history. Nevertheless, they are both emblematic of the villagers' engagement with the powers that control the water flow and the metabolic process they set in motion to access water. On the one hand, the mummy embodies the cosmopolitical order that defines humans' asymmetrical relationship with the mountains and nonhuman agents. In this hierarchy the former are submitted to the latter's good will, and to secure a stable water supply they must appease the mountains by bringing offering gifts, including the ultimate gift: a member of their own community.[28] Though valued as the most precious gift they could give,

Juanita, in the eyes of the villagers' ancestors, was simply one of many objects they offered to smooth the mountains' water metabolism. But unlike other offering items, Juanita stayed intact for five hundred years, providing the villagers with a link to their glorious but brutal pre-Hispanic past.

In fact, mountains played a key role in the villagers' cosmology right up to 1983. As Cabanaconde's main water supplier, Hualca Hualca was particularly important. In preconquest Peru the villagers were called the children of Hualca Hualca, which literally gave them an identity as their heads were deformed to imitate the shape of the mountain (Gelles 2000, 29).[29] The villagers' attachment to Hualca Hualca was confirmed during Cabanaconde's annual offering ritual. Elderly villagers relate that the entire community used to walk up to the top of the mountain at 6,025 meters (and during years of drought they would go twice) to make offerings called *q´apas* (or *irantas* as they are called in Tapay) to the deities; every year some of them remained at the foot of Hualca Hualca. Gelles reports that during the three days the ceremony lasted some of the villagers "would climb to the snow of the mountain itself at around 17,000 feet. Here they had to cut a central channel through the snow to redirect the snow melt to the main channel" (2000, 57). Gelles continues: "Many people were known to faint, become ill, and even die" (2000, 57), adding that they stopped doing it after making the hole in the Majes channel in 1983 (2000, 57). According to Gelles, the villagers remained at the foot of Hualca Hualca for three days, where they cut a channel through the snow to redirect the snowmelt to the main canal and thus increase the flow of water (2000, 57). But he also writes: "Ironically, the courageous effort to open the Majes Channel, which required great communal unity, brought to an end the yearly sojourn—part work party, part pilgrimage—to Hualca Hualca, and with it an end to the annually renewed solidarity that it provided" (2000, 58). When I interviewed an old man in 2011 I asked him if the villagers still believe they are descended from Hualca Hualca. He laughed and said: "No, that's a long time ago." In other words, Cabanaconde no longer recognizes the fictive kin bond that formerly tied it to the mountain.

In very concrete terms the 1983 action also aimed to speed up water metabolism. Using first manpower and later dynamite it redirected loads of water from the channel into the Hualca Hualca River. And as the state yielded to Cabanaconde's claim and included it in the Majes Project, the villagers' worldview altered from a cosmopolitical order dominated by nonhuman agents that grant the villagers access to natural resources such as water and land in return for offerings gifts to a realpolitical order established by the state that demands taxes, tariffs, and tickets in exchange for public services (roads, schools, sewers, electricity, health posts, water infrastructure, etc.) and a water supply that is both more stable and more abundant than Hualca Hualca. Rebellion ended up in cooperation and a shift in worldviews. Yet even though Cabanaconde stopped its collective offering rituals to Hualca Hualca and adopted a water management model designed by the state, its underlying understanding of water as a good that must be accounted for remained unchanged. Rather than only accounting to one power (Hualca Hualca), the villagers now do it to several powers (the mountain, the water committee, the state, and the water user association), though by different means (offerings to the first and tickets, taxes, and tariffs to the rest).

Apart from feeding an irrigated area called Joyas, Hualca Hualca still supplements the channel's supply in La Campiña, and although some villagers pointed out to me that the mountain's contribution today is insignificant, in 2011 the president of La Campiña's water committee told me that he had hired a *paqu* to climb halfway up Hualca Hualca to make an offering to the mountain. In a similar vein, some water users continue to make individual *pagos* to the canals and offtakes that feed their fields even though they recognize that they disagree on whether the offering is directed to Hualca Hualca or the channel. Following a ritual calendar similar to Tapay's offerings, the presidents of Cabanaconde's irrigation committees pay annual tributes to its five valves. Starting on August 1 at Tomanta, which feeds La Campiña, tributes are paid to Joyas on August 5, Villa de Colca (also called 18) and Castro Pampa on August 8, and finally Media Luna on August 15. The offerings' conflation of water suppliers (the mountain as well as the channel) and convergence with cash payments to different water providers (tax to the state and tickets to the water committees) and service agent (tariffs to the user association) suggest that unlike Tapay, which continues to make offerings to Seprigina while opposing the tax and the tariff, Cabanaconde views them as complementary. To put it differently, engaging in contractual metabolism with the state does not necessarily lead to a break with existential metabolism with the *apus*.[30]

CLIMATIC DISCORD

When I ask the villagers of Cabanaconde what people praise most about the district, their reply is its corn, particularly when prepared as *qancha*

(Quechua: roasted dry corn kernels). As one villager pointed out to me, "our *qanchas* are like crackers," a good demanded by everybody. Sadly, climate change is endangering Cabanaconde's corn production. Since 2011 villagers have showed me how worms are eating the corncobs and bugs are destroying the corn husks, attributing these intruders to rising temperatures. Equally, they say that there are more birds like doves eating their crops. Others complain that the sun burns more than it used to and that their crops therefore need more water. Yet others say that the rain is more irregular and that droughts are more common. And just about everybody agrees that the ice on Hualca Hualca is melting and the mountain produces less meltwater than before. Seemingly climate change is having serious impacts on the villagers' livelihoods. To document their climate perceptions and view of the future in a world suffering from rapid environmental change, I conducted a survey in Cabanaconde in 2012, revisiting the same households I had interviewed in my 2011 survey discussed earlier (Paerregaard 2018a).

My climate survey reveals a unanimous consensus that climate change is real in Cabanaconde; a few of those interviewed even asserted that the climate has changed radically in their lifetimes. A vast majority affirm that the drinking water supply has diminished in the last two decades, while almost one-quarter find that it remains the same and only two people state that it has increased.[31] In terms of the effects that climate change has on the villagers' lives, several of those interviewed gave more than one reply. Slightly more than half of those interviewed state that their health suffers. Many say that their skin is more exposed to the sun, and several also complain that they are afraid of cancer and viral diseases. One even contends that her blood pressure has gone up because of climate change. In one-third of the replies, those interviewed point to the economic effects that climate change has on farming and animal raising, which they claim are less productive than before due to irregular rainfall, frost, new crop diseases, and other environmental changes. Also, in one-fifth of the replies complaints are made that the temperatures are more extreme now, making the days too warm and the nights too cold. One of those interviewed did not offer any explanation for climate change. The causes of climate change can be many if we are to believe those interviewed, who in many cases provide not just one but several replies. Of these three-quarters claim that the cause of climate change is the villagers' own contaminating activities, such as sewage outlets, the burning of waste, tree cutting, the use of chemical products, the disposal of plastic items, and the displaying of fireworks during Cabanaconde's annual fiestas. A female villager even asserts that her use of perfume is causing climate change. Only a small percentage of those interviewed locate this outside Cabanaconde, pointing to Peru's mining industry and the explosive number of cars in the country's cities, while even fewer of the replies refer to global warming and to the hole in the ozone shield. One interviewee claims that it is the nuclear tests of the world's big powers that trigger climatic change, while another says it is Peru's many earthquakes. As to solutions to climate change, some suggest more reforestation and organic production; a stop to contaminating activities in Cabanaconde, including tree and trash burning, more recycling, and better use of natural resources; more information and better education of the villagers; and better control of their use of chemical products. Finally, as many as three-quarters state that there will be less water in the future, some even claiming that it's not likely humans will survive the growing water scarcity and providing answers such as "we'll all die" and "humanity will not survive." A small group also anticipates more conflicts and wars in the world as a consequence of climate change.

Summing up, from my survey it is evident that (1) there is consensus among the villagers that climate change is seriously impacting the environment in Cabanaconde; (2) the vast majority of the villagers think that the drinking water supply in the district is diminishing; (3) many villagers are concerned about the effects that climate change has on their health and their livelihoods; (4) there is a general agreement that it is the villagers' own activities that cause climate change; (5) very few villagers locate the cause of climate change in other places in Peru or outside the country; (6) the villagers overwhelmingly agree that the impact of climate change must be mitigated by local initiatives such as reforestation, recycling, and the improvement of the villagers' education; and (7) the majority worry about the future water situation in Cabanaconde, and some even think that climate change jeopardizes life in not only the district but also other places in the world.

Interestingly, there are few variations in how the interviewees explain climate change and its consequences and how they envision the future. Thus, age and gender have little importance for the replies, except that those who said that climate change is caused by national or global forces such as the mining industry and nuclear tests were males, and those who said it was their own consumption of personal items such as perfume were women. The answers' uniformity is evidence that the villagers by and large (old as well as young, women as well as men) have adopted the global discourse on climate change, which they now use to account for the environmental changes they are experiencing. Yet as my conversations with the villagers show, few make sense of the terms associated with this discourse. Thus, when asked what causes such phenomena as global warming and holes in the ozone shield, many simply replied, "it's because of climate change" (*es por el cambio climático*) or "it's because the climate has gone mad" (*es porque la clima* [el clima] *ha vuelto loco*). This could also be viewed as proof that the villagers recognize the gravity of the situation and indeed are aware of the need to take action, even though few try to live up to their own suggestions for how to solve Cabanaconde's environmental problems.

The contrast between the villagers' recognition of climate change as real and their apparent lack of action to address it reveals a major challenge in Cabanaconde's efforts to adapt to the changing environment. By watching and reading the news and engaging in contact with migrants and tourists who visit Cabanaconde, the villagers have become adept in employing the glossary of the global discourse on climate change (global warming, holes in the ozone shield, smog, contamination, recycling, etc.). Even so, the majority attribute climatic change and the water scarcity it causes to their own agency and suggest mitigation rather than adaptation as the appropriate response when asked how to meet future challenges. While some mention trash burning, the use of perfume, and the burning of fireworks, others point to modern chemicals as the cause of climate change, disregarding the fact that most villagers use farmyard manure. Rather than referring to the developed world when identifying the causes of global warming they blame themselves; likewise, rather than suggesting national or global solutions to solve its problems, they believe the issue should be addressed locally through such measures as recycling, reforestation, and better education. In other words, the villagers believe that the answer to climatic change is to moderate their own newly acquired modern lifestyle. Finally, the disillusion and despair prevailing in the answers to the last question of the second survey ("we'll all die," "humanity will not survive," etc.) suggest that many villagers feel abandoned and helpless when imagining the future world. The answers bring to the fore a dire paradox: the villagers belong to Peru's poorest population sector and are among the most climate vulnerable in the world, yet they attribute climate change to their own agency and worry about its impact on the world's future.

Preparing for climate adaptation is a demanding endeavor for people any-where in the world, but the task is particularly challenging for those who live in marginal rural areas and whose primary wish is development, which in the conventional understanding implies more, not less, CO_2 emissions. In fact, rather than changing the lives of the poor for the better as traditional development work does, adapting to climate change prepares people for the worst and helps them make the best out of a situation they would rather be without (Adger 2006; Adger et al. 2003). Cabanaconde's confounding of climate mitigation and adaptation and reluctance to prepare for what is coming illuminate the difficulties of grasping and dealing with the trade-offs implied in climate change adaptation. The list of recommended actions that Cabanaconde could take includes strengthening its general sustainability by introducing environmental protection of its watershed, new farming prac-tices that discourage tree burning and the use of chemical manure, modern-izing the sewage system and waste and recycling management, implementing a sustainable and fair drinking water system with high water tariffs in tourist hotels and low water tariffs in private homes, and more generally, bolstering Cabanaconde's own sense of agency and awareness of future environmental challenges. These activities are not likely to reduce Cabanaconde's vulnera-bility to increased water scarcity caused by climate change, but they will help it make better use of whatever water is available.

While these measures are preventive and may contribute to environmental sustainability in the long run, the villagers' most urgent concerns are their livelihoods and, as described earlier, the microclimatic impact in the form of new bugs and plant diseases that heat, irregular precipitation, and water scar-city have on Cabanaconde's agriculture, in particular the production of corn. The most effective way of counteracting these threats to the villagers' liveli-hoods is either to introduce new crops or to shift to fruit or vegetable produc-tion (as Tapay hopes to do) or livestock breeding (cattle, guinea pigs, etc.). However, as such changes demand not only investment capital and expertise on behalf of the individual farmers but also an overhaul of Cabanaconde's water management, which is organized to meet the needs of corn produc-tion, they require the support of district authorities and water committees and, even more important, the approval of the water users. As discussed pre-viously, even though Cabanaconde changed its irrigation model in the 1990s and replaced its own community model (using appointed *controladores* and the dual competition method) with a state model (using remunerated water allocators and the *de canto* method), water management still follows a uniform scheme that allocates water in four rounds during the dry season indiscriminately regarding variations of crops and actual needs. It is only in years of drought that the water allocators depart from the established irrigation scheme between January and April and in return for money allocate water in accordance with the water users' individual demands. Still, Cabanaconde only employs the *de canto* method, which leaves the water users waiting for their turn as the *controlador* allocates water from the highest to the lowest located fields, not the *brinco* method, as Tapay sometimes does, which allows the water allocator to direct water sequentially to the several fields of the same water user.[32] Another underlying problem in Cabanaconde's water management (as well as in Tapay and the rest of the Colca districts) is its use of surface watering: irrigation by flooding the field and distributing water over the soil surface by gravity, which leads to waste of water as it either evaporates in the sun (when irrigating during the daytime) or spills over the fields, flooding pathways, streets, and roads.[33]

The road map for Cabanaconde's climate adaptation therefore has three essential yardsticks: introducing new crops and forms of production, adjusting irrigation practices to individual needs, and making better use of the water supply. Many villagers are still hesitant to accept such changes, but a few have taken action. The following vignette illustrates the trouble Juan, who was president of La Campiña's irrigation committee when I got to know him in 2016, had to deal with when he tried to reform Cabanaconde's water management and adapt it to the needs of new crops and forms of production. Although Juan had verbalized his critique of Cabanaconde's irrigation practice at community meetings several times, his initial proposal was not to change them but to enhance La Campiña's water discharge by recapturing the water that goes waste from the Hualca Hualca River. The plan, Juan told me, was that once more water starts to flow, the water users of La Campiña would be persuaded to grant its water committee and *controladores* more room for maneuvering to manage water. Interestingly, Juan's action sheds light on not only the frictions that Cabanaconde's efforts to adapt to climate change cause among the villagers but also the implications they have for their notion of the state and, of particular relevance for the topic of this book, water metabolism and the role that offerings play in smoothing it.

In 2016 Juan invited Lardi, an engineer working for the ALA in Pedregal, to help him estimate Hualca Hualca's water discharge. Juan, who already knew Lardi and who had requested his visit to Cabanaconde as a personal favor, intended to use his estimates as proof of Hualca Hualca's continued importance for Cabanaconde's water supply and as support for his efforts to restore and reopen the waterway that directs the mountain's meltwater into La Campiña's infrastructure. Since Cabanaconde gained access to the Majes channel, La Campiña's water users have neglected the maintenance of the waterway, which today supplements only Joyas's water supply. It was Juan's hope that recapturing Hualca Hualca's meltwater would win him the support of La Campiña's water users, but he also knew that the project could be an invitation for trouble. In the world of water, one's win is another's loss, and no one expected Joyas's water users to let Juan take their water without a fight. Juan's plan was further complicated by the water users' divided interests. Some water users in La Campiña also have fields in Joyas, where they have been granted land as community members. At the same time, some of Joyas's water users have no stake in La Campiña. As outsiders they do not own land in Cabanaconde but rent or sharecrop the fields of native villagers who live in Lima or Washington, D.C., and they feel little need of adapting to climate change. Arguably, Juan's project was like opening a can of worms.

But Juan was in a hurry. His term as president of La Campiña's water committee was about to expire, and with only one general assembly remaining to present proof of the potential contribution of Hualca Hualca's waterway to its irrigation, Lardi's visit was crucial to making his case. Juan could of course have made the estimates of Hualca Hualca's water flow on his own, but as an ALA engineer Lardi represents the state and embodies Western science and the modern concept of water, which was critical for Juan's argument that Cabanaconde needs to reform its irrigation management and change crop rotation to adapt to climate change. Explaining the need to adapt, Juan told me: "Corn takes eight months to ripe and is too vulnerable to the rising temperatures. We must change crops to potatoes, peas, and alfalfa that only take two to four months to ripe." But to do this, Juna pointed out, La Campiña needs more water. "That's why we still depend on Hualca Hualca. Too much water is lost here. They say Hualca Hualca produces 200 liters [per second] and that Joyas takes 55 liters but we capture less than 40. Most of the water is lost on the way down to La Campiña. If we construct a canal, we could save a lot of it." It was Juan's hope that Lardi's numbers would help gain the water users' support for his project and calm their concerns that changing crops causes more water shortage and that recapturing Hualca Hualca's waterway would reignite their incipient water conflict with Joyas's water users.

Lardi arrived in Cabanaconde at 11:00 a.m. and took off on an empty stomach in Juan's car, together with two other members of the water committee (and me). After an hour's drive and two hours' walk, we stopped at 4,200 meters in a gorge that Lardi found suitable to make the first of several planned estimates of Hualca Hualca's water discharge. Before conducting these, however, Juan called upon us to take part in an offering ritual to Hualca Hualca that he had prepared to ask for the mountain's permission. Following the same script described in chapters 1 and 1, Juan arranged a *q'apas* that included the burning of offering gifts, saying prayers to the mountain, drinking alcohol, and chewing coca leaves. After the ritual, Juan took off his shoes, stepped out in the ice-cold water, and applied the measurement stick Lardi had passed him to estimate the discharge. Afterward, Lardi wrote down the numbers in his notebook while Juan asked me to take photos of the measurement stick. "Better have Lardi and you presenting the numbers than me. People don't trust me." Later, Lardi and Juan repeated the exercise twice farther down the mountain, first at 3,900 meters, right above Joyas's offtake, and later at 3,700 meters, just below it. Lardi's data showed that while Hualca Hualca still produces a considerable amount of meltwater, Joyas captures most of it, leaving almost nothing for La Campiña. Triumphing Juan said, "Look at the numbers; my guess about Hualca Hualca's water discharge was wrong. Joyas takes almost all our water. These numbers show that I was right and that we can claim our share of the water back." During the trip back to Cabanaconde, Juan affirmed his gratitude to Lardi, inviting him and me to his grandchild's birthday party in the evening. Lardi and I accepted Juan's invitation, pointing out that his visit was driven as much by a personal interest in Cabanaconde's affairs as by work obligations. Nonetheless, we never made it to the party. While drinking a couple of cold beers together upon our return to Cabanaconde, Lardi said to me: "We better not go. I know myself. I'll get drunk and then I don't make it back to Pedregal for work tomorrow."

When I returned to Cabanaconde in 2017 Juan told me that his project still had a long way to go. Lardi's numbers had impressed some of the water users at the assembly he led in 2016 as president of La Campiña's water committee, but the fear of triggering a new conflict with Joyas's water users and concern about all the work it would require to restore Hualca Hualca's waterway and construct a canal to connect it to La Campiña's irrigation infrastructure made many reluctant to support the project. Juan said: "People are lazy and don't like hard work. It's true the canal will take many hours of *faena* [communal work] but we will gain a lot. I'll keep on pushing until we get it done." Juan's story illustrates the obstacles innovative villagers who understand the need to adapt to climate change encounter when trying to change Cabanaconde's water management and introduce new crops and forms of production.[34] To many villagers, corn epitomizes Cabanaconde's history and culture just as water management is a central element in its community identity. Even though they recognize that climate change represents a threat to their livelihoods, adapting to its consequences implies difficult trade-offs that few are willing to accept. But Juan's story is also a window on how culture can contribute to climate adaptation by creating a feeling of continuity and reinforcing the villagers' sense of accountability as they replace water suppliers. At a time of rapid change, environmentally as well as politically, offerings represent a means to communicate and reassure their alliances with not only old power holders (mountains and ancestors) but also new power players (the state, water associations, etc.). As Juan demonstrated on the trip with Lardi to Hualca Hualca's waterway, paying tribute to the mountain and the channel go hand in hand. They both smooth water metabolism.

WATER CLAIMS

Water shapes Cabanaconde's history in remarkable ways. Five hundred years ago the villagers sacrificed the most valuable thing humans can offer: a young productive and reproductive member of their community. The sacrifice was a tribute to Ampato, which supplied Cabanaconde with water in preconquest times. After Cabanaconde gained access to a state-built channel four decades ago the villagers stopped making collective offerings to Hualca Hualca, which had replaced Ampato as its principal water supplier. To understand this change in ritual practice we must focus on Cabanaconde's idea of the powers that control the water flow and the relation of accountability humans engage in with these powers to induce them to produce and release water. The Majes Project was a game changer in Cabanaconde's perception of water and power. It altered the villagers' image of the state and made them change not only their offering practice but also their water management and prompted them to remunerate their water allocators and pay water tax and water tariffs. However, Cabanaconde hasn't abandoned its offerings altogether. The presidents of the water committees pay annual tributes to the channel's valves and, as Juan demonstrated, some of them even do it to Hualca Hualca on special occasions. Some of the water users also make offerings to the offtakes and canal intersections on an individual basis. In short, to the villagers, paying tribute to the mountain, tax to the state, and tariff to the user organization and buying the ticket from the water committee are not mutually exclusive practices. Rather, they are complementary ways of accounting for their water consumption to their shifting water suppliers.

Cabanaconde offers an interesting case to anticipate what may happen in other Andean communities that suffer from climate change and its impacts on their water resources. Like Cabanaconde, they have struggled with water scarcity for many years, but out-migration and demographic decline have alleviated the demand for water. More recently, however, glacier retreat, snowmelt, and irregular precipitation combined with rising temperatures and changing microclimates have created an unprecedented water crisis in many parts of the Andes, forcing Andean water users to look for new and untried strategies to access water. One strategy is to turn to the state and by legal or extralegal means obtain rights to its hydraulic infrastructures, as has occurred in Cabanaconde. Another is to gain the support of private agents, as is currently happening in Tapay. The case study of Cabanaconde suggests that when Andean water users shift to the state as their new water supplier, offerings continue to play a critical role in their relation of accountability with not only the mountains but also the state. Tapay will show whether the same happens when they shift to Peru's mining companies.

FOUR

Huaytapallana

THE *APU* THAT IS DYING

The sun was gaining strength, and the morning noise from Huancayo's traffic signaled that the city was heading for another busy day. It was June 21, 2014, and I was waiting for Carlos, a Peruvian anthropologist and an acquaintance of mine for many years, who had promised to pick me up for a daylong ride to attend an offering ceremony at Huaytapallana, the third highest mountain in Peru's central highlands, located just outside the city of Huancayo. At 8:00 Carlos turned up on a motorbike, along with his son, Rolando, who was driving a four-wheel-drive sports utility car. On Rolando's invitation, I entered the car, joining the three other passengers, and shortly afterward we were making our way out of Huancayo, a regional hub for trade, smaller industries, education, and other economic activities in Peru's central highlands. Once we had left Huancayo the road changed from paved to dirt and got narrower, which made the climb not only more strenuous but also more congested. During the week, it is mostly truck drivers who use the road to bring merchandise to and from Huancayo, but on this day the traffic included private cars and buses packed with people of all ages who all were going to Huaytapallana, the mountain described in this book's first chapter.

Our only stop between Huancayo and Huaytapallana was the community of Acopalca, situated at the edge of the *puna*, the bleak Andean upland above 4,000 meters. Officials from the regional government of Junín had set up a roadblock in the community to inspect the passing vehicles and register the passengers. They also asked travelers to sign a declaration affirming their intention to protect the environment of Huaytapallana in accordance with the decree the Peruvian government had issued in 2011, which declared the mountain an area of protected conservation and seeks to mitigate the environmental impact on its melting glacier. Rolando signed the declaration and returned it to the inspectors, who after a quick look at me and the other passengers told us to move on.

Shortly afterward we entered the *puna*, where an unsettling scenario was taking place: a long queue of vehicles struggling to make the last steep rise to Huaytapallana produced a thick smoke that disturbed the view of the mountain's white glacier. The image of polluted mountain air generated a heated discussion among my co-passengers about the visitors' behavior and the authorities' lack of control. Our outrage at other people's carelessness only got worse when we saw the plastic plates, cups, and bottles and other kinds of litter that previous visitors had dropped along the road. One of my co-passengers asked: "How can people behave like that? Imagine the damage they do to the environment!" Another declared: "Visitors who leave their garbage should be fined. Or even better. They should be banned from coming here." We continued discussing different ways of alleviating the impact the growing tourism has on Huaytapallana for a while until the third co-passenger rhetorically asked: "Well, we're all part of the problem, aren't we?" Her question brought the conversation to a sudden stop. Then Rolando cried: "We're here. Isn't it beautiful!" In front of us Huaytapallana's ice-covered summit appeared between the blue sky and a turquoise-colored lake (see figure 7).

The ride in Rolando's car was my first of three field trips to attend the offering ceremony at Huaytapallana on June 21. Between 2014 and 2017 I had the opportunity to observe how the relationship between the event's organizers and their followers, on the one hand, and the regional authorities and other local actors, on the other, have developed, and how the very idea of making offerings has become an issue of contestation because of climate change and the retreat of Huaytapallana's glacier and the visitors' environmental impact on the mountain.

Having conducted fieldwork in the Huancayo area in the early 1980s, for me Huaytapallana's majestic beauty was no surprise. Studying the pilgrimage to the mountain's glacier, however, was a new experience, and even though the offerings to its deity reminded me of what I had seen in Tapay and Cabanaconde, they opened a new window onto the relationship of exchange that the giver and the receiver of the offering gift engage in. Labeling the offerings transactional metabolism, I explore them as a give-and-take relationship between humans and nonhumans that allows the former to ask the latter for specific goods and favor in return for offering gifts. Unlike existential and contractual metabolism, which are both ritual enactments of a water flow on

FIGURE 7. The glacier of Mt. Huaytapallana, in the central highlands of Peru, 2014. Photo by author.

which the lives and livelihoods of the communities practicing them rely, the physical need for water is not the main driver of transactional metabolism. Yet even though the pilgrims of Huaytapallana attribute to the mountain symbolic rather than material meaning, many are aware that its glacier constitutes their principal supply of drinking water. As in the previous chapters, I begin by introducing the ritual's historical and social context.

SHAMANS AND PILGRIMS

For years it was only local livestock raisers in Acopalca and neighboring communities that made offerings to Huaytapallana. Celebrating Santiago, the titular saint of llamas, alpacas, and sheep, on July 25 the villagers asked the mountain for permission to use its pastures (Altamirano 2014, 149–152).[1] Three decades ago, the mountain became the stage of another religious gathering that takes place on June 21, as described in the opening scene. The event attracts people from mostly urban areas who visit Huaytapallana for a variety of reasons, one being concern for the environment and the mountain's future, another the search for new (or old) social identities. The event dates to 1994 when an Andean shaman, Pedro Marticorena, gathered a small group of people to celebrate the Andean New Year and pay respect to Huaytapallana. The shaman's initiative was backed by some of Huancayo's urban intellectuals, who saw the event as a platform to revive the city's ethnic identity as Huanka (Huancayo's pre-Hispanic ancestors) in the wake of the political violence that had tormented it (and the rest of Peru) in the late 1980s and early 1990s. The attendance remained small for several years, but in the mid-2000s it increased when Huancayo experienced rapid economic growth and the introduction of modern consumer habits and lifestyles. Huancayo's new prosperity, however, has also created environmental problems, social insecurity, and political corruption, inducing many people to question the meaning of modernity and turn toward the city's shamans and its ethnic past in the search for moral guidelines and spiritual inspiration to conduct their lives.

When Pedro and his supporters began celebrating the Andean New Year at Huaytapallana in 1994, Peru was recovering after a decade of economic recession and political violence. The country's crisis peaked in the late 1980s when inflation and poverty rates reached unprecedented heights, and the Peruvian military and the Shining Path, a revolutionary organization, became entrenched in a savage war that caused sixty-five thousand deaths, the majority of whom were civilians, and forced thousands of others to take refuge in Lima, Huancayo, and other cities in Peru (Gorriti 1999; Starn 1999). The conflict peaked in 1992, the same year security forces captured the Shining Path's leader. Two years earlier, the newly elected president, Alberto Fujimori, had introduced a neoliberal program that deepened the country's economic crisis and triggered massive emigration. Even though many disapproved of Fujimori's economic policy, they gave him credit for the defeat of the Shining Path. Moreover, to alleviate the growing poverty Fujimori launched a series of social programs that he used to generate a popular movement in favor of his government and break the ground for new class alliances.

In Huancayo, which was one of the areas mostly affected by the violence that haunted Peru (Paerregaard 2002), the economic and political crisis gave rise to ethnic awareness and paved the way for the election of a new mayor who promised to restore people's belief in the future by strengthening their Huanka identity. Among the mayor's many initiatives was the construction of El Parque de Identidad, a park that exhibits Huanka culture in the form of sculptural and architectural installations and is today one of Huancayo's hallmarks. During his nine years in power, the mayor also supported a Huanka revival movement that was started by a group of urban intellectuals who sought inspiration in the same Andean cosmovision Pedro and Carlos (introduced in the opening scene) subscribe to, and that gained momentum in the postconflict era. The movement continued to grow in the 2000s, and even though the reclaiming of the region's ethnic identity still ranks high on its agenda, this now includes a new issue: Huancayo's environmental problems and the uncertainty they create about the city's future.

The pilgrimage to Huaytapallana now exceeds one thousand and appears in the headlines of Huancayo's newspapers every year in June; it has become a topic of debate in the city both because of its spectacular ritual and the environmental problems this causes. An important driver behind the pilgrimage's surge is the search for new identities and new meanings in modern life, which my conversation with the co-passengers in Rolando' car reveals. One of them told me she was born in Huancayo but had lived most of her life elsewhere, including several years in the United States. Nonetheless, she affirmed, "I feel very Peruvian!" a feeling she said had grown stronger upon her return to Huancayo a few years earlier. She said this was her first trip to the mountain and explained, "By going to Huaytapallana I express my feeling of being Peruvian," adding, "but it also makes me feel Huanka, so Huanka." Another passenger said that it was also her first trip to Huaytapallana. The woman claimed that she wanted to pay respect as this could bring her luck. "If I give the mountain something it listens to my wishes and returns my offering," she asserted. The third woman, on the other hand, told us that she has been on Huaytapallana several times, though never on June 21. She said that it was the celebration of the solstice and the welcoming of the Andean New Year that had incited her to visit Huaytapallana on this specific day. One of the women then inquired into my reasons for visiting Huaytapallana. "Curiosity," I replied. And when the woman asked Rolando the same question, he responded: "I live off tourism and I go there whenever someone pays me. Sometimes I spend the night there with a group of tourists, so I know the mountain at day as well as at night." Even though my travel companions and I were attending the same event, then, we were driven by very different motives.

I returned to Huaytapallana on June 21 in 2015, this time with Carlos as driver. Interestingly, the conversation in the car revolved around the same issues as in 2014, even though my travel companions were new. Among these was a middle-aged woman traveling with her teenage son. The woman opened the conversation by telling us that this was her first trip to Huaytapallana and that she had brought her son so he could learn how to respect the mountain. "I sent my children to a school that values children's individual qualities and teaches them to be independent and take responsibility. The school also organizes tours outside Huancayo to teach the children how to prepare their own food and take care of themselves." The woman also said she deplores the consumer habits of today's young people, who in her view eat unhealthy food and spend too much time watching television and playing video games. Visiting Huaytapallana and participating in the offering ceremony on June 21, the woman asserted, provides a new perspective on life and allows "you to reencounter yourself in a more authentic environment." My third travel companion was a young woman who affirmed that she too wanted to "reestablish the balance in my life" by making contact with the mountain. The woman told us that she had left her two small children with their grandparents but that she hoped to take them to Huaytapallana when they grew up.

A common theme in the discussions with travel companions on my trips to Huaytapallana has been their quest for alternative perspectives on Huancayo's recent modernization and inspiration to fuel their everyday lives with new energy. Examples of this search were my co-passengers' affirmation of national and indigenous identities and their efforts to escape the city and establish contact with nature. By the same token, many of the visitors I talked with on both trips conceived of the mountain as a spirit with the capacity to influence its own and others' lives, and saw the offering as a means of tapping into its power to gain political capital, make economic profit, attain public notoriety, or conversely, put a spell on someone they wanted to harm. More bluntly, they all tried to capitalize on Huaytapallana and its glacier, which yields life to the city by supplying it with fresh water and therefore symbolizes not only the continuity between its ethnic past and modern present but also the energy that will assure the life of its future generations.

At the same time, visiting Huaytapallana was an occasion for concern. Years of environmental pollution have taught Huancayo's inhabitants that economic growth and modernity come at a price and that natural resources such as water are scarce and vulnerable to contamination. The visit reminded many that the mountain is no exception to this development and offered them a firsthand view of how humans impact nature. More importantly, they saw the environmental consequences of their own and other visitors' activities, which many interpret as the real cause of glacier melting. As described in the chapter opening, the disturbing view of car pollution and nonorganic trash on the road to the offering site induced some of my travel companions not only to suggest stricter control of people's access to Huaytapallana but also to question their own presence on the mountain. The visitors' concern for the environment was complicated by their own part in the June 21 event, which exposes an inherent dilemma in modern-day offerings between, on the one hand, offering gifts to the nonhuman agents and thus contributing to their well-being, and on the other, abandoning the offering items on the ceremony site and thus polluting the environment.

DEPLETING WATER

Huaytapallana is emblematic of Huancayo's environmental problems in various ways, but one is of particular concern. The mountain's glacier constitutes a critical water source for approximately 500,000 people, of whom 88 percent live in the city and 12 percent in the rural surroundings (Milan and Ho 2014). To the east its meltwater feeds a watershed that flows toward Peru's jungle and the Amazon basin. And to the west it runs into the Shullcas River, which supplies a range of water users. The first to use its water are the alpaca and sheep herders of the community of Acopalca located at 3,900 meters, who use the pastures above 4,000 meters to graze their animals. Farther down the communities of Chamisería, Pañaspampa, Vilcacoto, Uñas, and Palián take water from the Sullcas River before it reaches Huancayo (3,259 m), with a population of 450,000. Once it has passed the city the river continues to feed several communities (Cochas, Cullpa, Aza, Racracalla, and Paccha) before reaching its outlet in the Mantaro River at 3,200 meters.

The main livelihoods of the Shullcas communities are agriculture, husbandry, commerce, small industry, and fish farming, which all require water and in many cases contribute to the increasing contamination of the Shullcas River and its environment (USAID and Catie 2016, 111). Rural-urban and in some places international migration is also widespread in the communities, helping people to complement the income from their conventional livelihoods.[2] Even though some farmers rely on rain, 72 percent of the area's fields are irrigated, which represents the major source of water consumption in the rural communities (2016, 124). Of all the water used for irrigation, 43 percent is tapped from the Shullcas River, while 34 percent comes from local wells or springs (2016, 125). In total, irrigation, trout farming, and a few other activities consume 50 percent of the Shullcas River's water. The pressure on the river's water discharge is exacerbated by contamination, which the bulk of the area's irrigation users think originates from mines, industry, or domestic use and which they believe constitutes a threat to their crops. Yet at the same time many say that they make use of chemical fertilizers, insecticides, or other substances in their agricultural production (2016, 125). Not surprisingly, water quality ranks as high as water scarcity on the communities' list of environmental concerns.

But the region's main challenge is not to ensure the water supply of agriculture and other rural and semi-urban activities such as small industry and fish farming, on which only a small percentage of its population depends. It is to provide Huancayo's inhabitants with clean water to drink, cook, and wash. A critical political issue is therefore the availability of water resources, which are increasingly under pressure due to population and economic growth, poor infrastructure, changing lifestyles, and climatic changes (López-Moreno et al. 2014). Since 2005 the city's water authorities has been forced to ration water, and it is expected that by 2030 there will be water shortages for one-third of its population, who will experience regular water rationing (Gómez and Santos 2012).[3] Viewed against the backdrop of Huancayo's other water sources, the prospect of supplying the city's growing population with water looks particularly gloomy. Thus, wells meet a large part of Huancayo's current water demand, but because underground water does not renew in the short or medium term, over time the city will become even more dependent on Huaytapallana's glacier lakes and neighboring water sources (López-Moreno et al. 2014).

To alleviate the pressure on Huancayo's underground water and enhance its water capacity, SEDAM, the region's public drinking water manager, which services 95 percent of the city's population, has started to take water from the headwaters (*cabezera*) of the Shullcas River's watershed, which includes ten lakes.[4] Of these lakes four are glacier lakes and six rain lakes fed by what is known as "water harvest." SEDAM has also begun to tap water from local springs and capture water that filters from the nearby slopes, and it has plans to build a dam close to Acopalca and to take water from Huaytapallana's wetland and other surface waters (a technique known as "water sowing").[5] Still, while these water supplies help diminish the demand on the city's underground water, their use will only deepen its dependence on Huaytapallana's glacier, which is the principal source of the glacier lakes and the Shullcas River's watershed, which feeds the wetland, the springs, and the slopes. Likewise, harvesting water by capturing the rain and storing it in lakes is just another way of taking water further up the hydraulic circle and thus reducing the water discharge of the Shullcas River's watershed. Eventually Huancayo will have to look for alternative water sources.[6]

To mitigate the competition over the Shullcas River's water and the social conflicts between different user interests (pastoralism, agriculture, fish farming, extractivism, tourism, etc.) and, not least important, to meet the growing demand for domestic use by Huancayo's urban population, ANA (see chapter 1) has taken steps to establish a so-called water basin council, similar to initiatives in other parts of Peru (such as the multisectoral committee I discussed in chapter 3). Aside from the many water users organized in Juntas de Usuarios, the council will include public service agencies such as SEDAM, JASS, regional industrial and mining companies, peasant communities, different expert groups, and ANA/ALA.[7] So far the plans are preliminary, and in the meantime ALA-Huancayo continues to negotiate the region's water crisis singlehanded.

The quality of the city's water supply makes up yet another environmental concern. Peru's Ministry of Energy and Mining has recently granted permission to a mining company to explore mining options in four places within Huaytapallana's protected area. In a region that has a long history of mining, this comes as bad news for many. For years discharge from the mines in nearby La Oroya has polluted the water of Mantaro, the biggest river of Peru's central highlands. The mining company responsible for the pollution has recently stopped operating due to pressure from a popular movement in Huancayo called Mesa de Diálogo Ambiemental Huancayo y Junín, but it will still take many years before the river recovers.[8] Moreover, even though traditional mining in the Huancayo region now has been brought under control, pollution from illegal exploitation of natural resources such as marble and sand in the Huaytapallana area, and also from local fish farming along the Shullcas River, endangers Huancayo's freshwater supply, demanding new and more consistent environmental policies.

TRANSACTIONAL METABOLISM

It was almost 11:00 a.m. when we reached Lazuntay (also called Lazu Huntay), one of Huaytapallana's four glacier lakes, where the offering ceremony was going to take place (see figure 8). Several hundred people had already

FIGURE 8. Pilgrims gathered at Lake Lazuntay at Mt. Huaytapallana, in the central highlands of Peru, 2014. Photo by author.

arrived, and in the following hour the number of participants increased to more than five hundred. The bulk of the visitors were located on the north side of the lake (see figure 8). They were mainly families and small groups of individuals from Huancayo and other cities in the region, many of whom were first-time visitors like myself and my travel companions. On the south side, around thirty members of a rural community from Huancavelica, a neighboring region, had gathered. Finally, a small group representing a regional political party called Bloque Popular, which aims to introduce socialist reforms inspired by the Inca society in Huancayo, had set up their small camp at the main entrance to the lake. Even though I visited the two latter groups, I spent most of the day on the north side of the lake in the company of Carlos and the people he introduced me to.

Many people had prepared *mesas* (Spanish: tables), which consist of pieces of woven cloth on which all the items brought for the mountain are placed. Among the objects to be offered were fresh and dried corn, flowers, a wide diversity of fruit, bread, biscuits, dried beans and lentils, sugar, chocolate, candies, fried chicken, plates of prepared food, herbs, *chicha* (Spanish: corn

FIGURE 9. Offering gifts to Mt. Huaytapallana, in the central highlands of Peru, 2014. Photo by author.

beer) and spirits in plastic bottles, beer, soda, wine, whisky (Chivas Regal in original cart boxes), cigarettes, coca leaves, and candles (see figure 9). Some *mesas* also included small figures of men and women engaged in sexual acts, toy cars and trucks, advertisements for electronic artifacts, and wallets with ID papers, which suggested that the givers expected the offering to yield a return in the form of fertility, wealth, or legal status. Other visitors shared the expectation that the ceremony brings good fortune. Thus, I saw a young couple engage in an informal marriage ceremony in front of the mountain. I also spoke to several people who said that they had come to request Huaytapallana's support to recover from illness or heal their physical incapacities. Driven by the hope that Huaytapallana's meltwater has miraculous power, some even took a swim in Lazuntay or let relatives throw water from the lake on their naked bodies (see figure 10).

Shortly after noon the organizers called people to assemble in a circle around the many mesas and offering gifts. Facing Huaytapallana's glacier, a spokesman introduced the event. In a mixture of Quechua and Spanish he pointed out the importance of living in harmony with the *apus* and *pachamama* as "our ancestors have taught us." He also said that people must pay respect to Huaytapallana because it provides them with water and therefore is essential for their future existence. The spokesman, who was assisted by two flute players and two men dressed in yellow and red frocks and wearing

FIGURE 10. Pilgrims bathing in Lake Lazuntay at Mt. Huaytapallana, in the central highlands of Peru, 2014. Photo by author.

masks of animals, then presented the ceremony's three speakers, of whom the first two were *layas*, a Quechua term for specialists conducting Andean offering ceremonies. The first *laya* to speak was Pedro Marticorena, who wore a long cotton gown partly covered by a woven cloth and with the head and shoulders covered by a white cloth and delivered his speech in Quechua (see figure 11). He praised Huaytapallana for its spiritual power and expressed his gratitude for its generosity, after which he called out the names of not only the most important Andean mountains but also the principal elements of Earth (*pachamama*, the oceans, etc.) and of the universe (the moon, the sun, etc.). The man concluded by requesting Huaytapallana to receive the many offerings people had brought. Wearing mundane clothing and dressed less spectacularly than the first speaker, the second *laya* also started by directly praising Huaytapallana's sacred power in Quechua. He then turned to the audience and in Spanish expressed regret regarding the glacier melt and the harm it inflicts on the mountain, using Quechua terms such as *yawarsonqo* (his heart is bleeding), *llaquikuyan* (he is sad), and *waqayan* (he is crying) to emphasize its suffering. The *laya* also reminded the listeners that June 21 is the Andean New Year, a time to connect with their forefathers and learn to live in harmony with the *apus* and *pachamama*.

After honoring Huaytapallana in Quechua the third speaker, who wore a poncho and a stiff hat, switched to Spanish, urging the listeners to collect

FIGURE 11. Shaman talking to Mt. Huaytapallana, in the central highlands of Peru, 2014. Photo by author.

all waste on the ground. He pointed out that the organizers had signed an agreement with the regional government that holds them responsible for keeping the protected area of Huaytapallana clean. The man said that the mountain's current suffering is due to "the big transnational companies that exploit and hurt the earth," and to people living in the cities "who do not always understand how their contaminating habits damage the environment." On the contrary, the speaker emphasized, the people who live in Andean communities and who worship Huaytapallana and *pachamama* know how to live in peace with the environment. Without Huaytapallana Huancayo will die, he contended. Finally, a woman addressed the crowd in a mixture of Quechua and Spanish, underscoring the importance of such Andean values as respect for the environment and harmony between humans and their surroundings. The four speeches were accompanied by a small orchestra that played Andean flute music while two men, one standing on a rock close to Lazuntay and the other on the mountainside above the crowd, alternately blew a *pututu* (Quechua: a big seashell used as an instrument in Andean ceremonies to make a hollow sound).

When the speeches were over, a man offered coca leaves that he had gathered from the mesas in a bowl, asking people to keep the leaves that were intact for the offering ceremony.[9] Once the bowl had passed around the circle of participants, the spokesman announced that the central moment

FIGURE 12. Pilgrims making an offering to Mt. Huaytapallana, in the central highlands of Peru, 2014. Photo by author.

of the ceremony had arrived: the greeting of Huaytapallana to welcome the Andean New Year. Looking toward the mountain, he raised his arms and said in Quechua: "Great *apu*, we have come here to greet you and show you respect." In response, the crowd imitated his gesture, saluting Huaytapallana while the two men with *pututos* sounded their instruments. The solemn atmosphere lasted several minutes, after which people started to embrace and wish each other a happy new year. The forceful noise of a piece of ice breaking off the glacier, however, interrupted the cheerful mood for a moment, reminding people not only of Huaytapallana's spiritual power but also of the physical power that global warming and glacier melt release. People reacted very differently to the sudden breaking up of Huaytapallana's glacier. A man next to me said, "Huaytapallana has heard our prayers," while a woman asked, "What will happen to Huaytapallana when the ice is gone?"

In the third and last part of the ceremony people carried their offering gifts to the slope between the ice and the lake, where they settled in small groups and started to prepare the act of offering (see figure 12). Meanwhile, a brass band of around ten musicians dressed in suits and ties stood above the offering scene and entertained the crowd. I gathered with my traveling companions, Carlos, and Rolando around a small fire that some of their acquaintances had started. To initiate the ceremony, Carlos lighted the candles and the incense and set fire to some of the offering items (e.g., coca leaves), leaving the non-flammable items (fruit, fresh corn, *chicha,* and spirits) on the ground. The fact that only two matches were needed to light the candles and the fire, Carlos later told me, was a sign that the mountain had heard his prayers. He then opened the wine and the whisky, poured it into some cups he had brought, and invited everybody to drink. Before each drink we all dripped some of the wine or whisky on the ground before saluting Huaytapallana. Afterward Carlos asked us to empty the bottles by pouring the wine in a straight line below the fire and the whisky in a circle around it. He also offered us cigarettes and coca leaves to protect us against dangerous spirits. Smoke was now coming up from other fires, and all around us people were standing with their arms lifted to greet Huaytapallana. Shortly afterward I saw several individuals returning from the glacier with chunks of ice. I asked a young boy who held an icicle in his glove-covered hands why he had come to Huaytapallana and what he was going to do with the ice (see figure 13). The boy, who wore a suit and tie, answered that his father was a musician in the brass brand and that he planned to take the ice with him to school on Monday. "It will bring luck to me and my class," he explained to me.

At around 2:00 p.m. the offering ceremony was over, and people began to descend from the mountainside. While some prepared for the trip home, others stayed to chat in informal groups. A few even took a late swim in the lake. Two women had put up small stalls selling hot food, while a group of five elderly men stood around three cases of beer drinking heavily but without drawing much attention. I used this opportunity to mingle with people, asking them about Huaytapallana and the ceremony, and as time was short, setting up a couple of interviews in Huancayo for the days to come. A woman said to me that she was very happy to attend the event, but that Huaytapallana's glacier retreat had taken her by surprise. As described in the book's introductory vignette, a man next to her suggested that the mountain was growing old, and that the disappearance of the ice will lead to its death and a *pachakuti.* "It's something that has occurred in the Andean world before," the man claimed. He also contended that "climate change is nothing new." Another woman I talked with said that she was concerned about the wishes she had made. "I hope Huaytapallana has listened to my prayers," she told me. A man in his late sixties who said that he had

FIGURE 13. Schoolboy carrying an icicle from Mt. Huaytapallana's glacier, in the central highlands of Peru, 2014. Photo by author.

participated in the offering ceremony since it was initiated in the mid-1990s saw the ceremony as a political statement. He declared: "I'm here because I'm *anti-minero, anti-maderero* and *anti-petrolero.*" That is, he is against the exploitation of minerals, wood, and oil, three essential export commodities in Peru's economy. I later learned that the man is a judge in Huancayo well known for his viewpoints on environmental contamination, neoliberalism, and Andean culture.

Just before 4:00 p.m. the sun disappeared, and Rolando told my travel companions and me that it was time to leave. On our way back to Huancayo we briefly exchanged experiences from the event. Then the day's fatigue overwhelmed us, and we stayed silent the rest of the trip. We entered Huancayo in the dark and became engulfed in the city's evening traffic with its relentless noise and bad smoke. Huaytapallana's sparkling ice now seemed like a remote dream.

CLIMATIC DISARRAY

When I did fieldwork in the early 1980s in the communities of the Cunas River watershed situated opposite the Shullcas River, the villagers always complained of water scarcity (Paerregaard 1987b). And when I later lived in Huancayo for six months writing up my field notes, drinking water and sanitation were issues of almost daily conversations I had with people. But even though water scarcity has been a chronic problem in Huancayo for a long time, climate change has made the situation worse, reminding its population of the city's vulnerability and the environmental problems modern urban life entails. Until recently, few city dwellers gave Huaytapallana much attention except on Sundays, when they go on picnics in Huancayo's rural surroundings and enjoy the picturesque view of its glittering glacier from afar. Due to climate change, however, Huaytapallana is now a hot topic among the city's inhabitants. As the mountain is rapidly losing its icecap and as water has become rationed in several parts of Huancayo, a new awareness of the mountain's vulnerability and the role it plays as the city's water tower has emerged.

In fact, Huancayo's population received several warnings of the consequences global warming may have for not only Huaytapallana's glacier but also the safety of their own lives more than thirty years ago. In 1969 the dike of Lazuntay broke because of a seismic event, and in 1990 a GLOF (glacier lake outburst flood) occurred due to the melting of the glacier and the breaking off of an ice block, which caused the water of Chuspicocha to overflow and the lake's dam to collapse. In both cases the water flooded the Shullcas River, destroying hundreds of houses on its way to Huancayo, where it caused many fatalities (López-Moreno et al. 2014, 2). However, it was only when Huancayo began to experience permanent water shortages in the 2000s that people understood the linkage between climate change, glacier retreat, and the city's water supply. The gravity of Huancayo's water crisis is evident from a recent study on the glaciers of Huaytapallana and other nearby mountains, which estimates that for the whole Cordillera Huaytapallana, 56 percent of the ice covering the surface has disappeared (2014, 5). In fact, the research shows that Huaytapallana's glacier retreat is not a new phenomenon. From 1984 to 2011 it decreased from 50.2 km^2 to 22.05 km^2, and in areas below 5,100 meters the loss was as much as 80 percent. Even the summit areas (above 5,000 m) lost ice cover in this period (López-Moreno et al. 2014, 5). And even though the study characterizes the period from 1998 to 2011 as "alternations of stability, slight increases, and slight decreases of glacier area," the future of the region's glaciers does not look promising. The researchers made the following prediction: "Based on our observations of glaciers in Huaytapallana during the study period, an increase of 1.2°C (predicted by the "best case" scenario) would lead to the disappearance or near disappearance of glaciers in all areas that we studied except in the Huaytapallana area." And as to the danger of future GLOFs, they write: "The formation of new lakes in unstable high mountain environments is likely to continue, and this may lead to increased risk of flooding, especially because of the very steep and crevassed nature of these glaciers" (2014, 10).

As the researchers behind the study conclude, the principal reason for the region's glacier retreat is global warming. Even so, climate change is not the only cause of Huancayo's water problems. Mining, fishing, and other exploitative activities also represent a growing threat to the city's freshwater supply, and so do Huaytapallana's many visitors, who leave large amounts of trash in the mountain area every year and who have become a cause of concern for Huancayo's environmentalists. The organizers of the June 21 ceremony are increasingly taking the criticism of its anthropogenic effects seriously, and during the event speakers recurrently appeal to the visitors to collect their disposable waste. Yet even though some participants are becoming aware of the environmental impact of their activities, as described in the opening scene, I saw people leaving nonorganic residue from the offering ceremony on the mountain and throwing offering disposals in the glacier lake. Although several of the participants I spoke with recognized that the abandoned mesas constitute an environmental problem, they told me not to worry. "The herders from the nearby communities always pick up the leftovers after we have left," one man pointed out.

To make the situation worse, rather than participating in the event Pedro organizes on June 21, many people contract *layas* to arrange individual offering ceremonies asking the mountain for fortune in their economic affairs, individual careers, and family activities. This has not only generated a constant flow of visitors to Huaytapallana throughout the year but has also created a commercial market for *layas* in Huancayo. While some of the new *layas* are linked to parties such as the Bloque Popular, which read the Andean cosmovision as a political ideology, others reach out to business and restaurant owners or singers and dancers, who hope the offerings will produce prosperity or fame.[10] The leftovers on Huaytapallana also indicate that many visitors view the offerings as a means of asking for specific favors—some tangible, others intangible. Quite a few leave *alasitas* (see chapter 1) on the mountain, while others leave half-burned black candles or used underwear and other belongings of people to whom they wish to cause harm. Finally, many leave the disposable trash from the picnics they hold when organizing offering ceremonies on the mountain. According to the regional government, the total amount of trash left in the protected area of Huaytapallana every year amounts to more than six tons.[11]

Huaytapallana's environmental decay has not gone unnoticed by the national authorities, and in 2011 the national government passed a decree declaring the mountain an area of regional protection, covering an area of 22,406.52 hectares.[12] The decree's aim is (1) to conserve and restore high-Andean ecosystems such as grasslands, wetlands, lakes, and rival ecosystems; (2) to promote sustainable tourism and other human uses that contribute to maintaining the protected area; (3) to improve the capacities of the regional government in protected area management; and (4) to develop environmental education among local people and to strengthen research in and facilitate access to information about the protected area (Haller and Córdova-Aguilar 2018). While the act represents an important step forward in protecting Huaytapallana's glacier, it also has its flip side in respect to two important activities: extractivism and tourism. Regarding the former, Haller and Córdova-Aguilar point out that the decree explicitly permits the use of renewable resources while restricting the exploitation of nonrenewable resources (2018, 60). In relation to the latter, however, the study by Arroyo Aliaga, Schulz, and Gurmendi Párraga (2012) should be a cause for concern. They found that the activity of *turismo vivencial* (Spanish: community tourism) in parts of Huaytapallana is one of the principal anthropogenic activities that accelerate the loss of its snow mass by generating negative impacts on the glacier system. As the community tourism referred to by the researchers includes the pilgrimage to Huaytapallana, the decree leaves the regional government that is in charge of managing the protected area walking a tightrope between safeguarding the environment, on the one hand, and on the other allowing continuous access to the mountain and the glacier by pilgrims and other visitors.

Describing this dilemma as a trade-off between spiritual practices and the provision of clean potable water, Haller and Córdova-Aguilar write: "Regularly, Andean shamans come up and pay tributes to the apu or mountain deity." They continue: "These practices meet with more and more criticism from the PA [Protected Area] authorities, because the waste produced by the so-called pagapus (offerings) contaminates the water resources that subsequently flow down to the peri-urban and urban areas" (2018, 61). To constrain the activities of shamans and the agents who organize tours to

Huaytapallana (like Rolando, Carlos's son), the regional government has begun to register them and instruct them on how to mitigate their environmental impact on Huaytapallana at public meetings. At a meeting in March 2015 about thirty of the city's more than two hundred *layas* and a few tourist agents turned up together with members from Mesa de Diálogo Ambiental Huancayo y Junín. The organization surveys the regional government's protection of Huaytapallana's environment and has helped it produce a master plan to preserve the mountain (Plan Maestro del Àrea de Conservación Regional Huaytapallana), which consists of a list of proposals to safeguard Huaytapallana's protected area, including a ban on bathing in the glacier lake and touching the ice. During an interview I conducted with the former director of the office responsible for Huancayo's environment in 2015, she said: "We need to control the growing number of visitors, and in particular, the waste of the offerings they make. We have asked the *layas* not to leave any trash, but they don't always listen."[13]

A member of Mesa de Diálogo Ambiental Huancayo y Junín (an anthropologist from Huancayo's university) whom I also interviewed in 2015 concurred with the director's worry. However, she pointed out that the new regional government that came to power in January 2015 has done little to implement the master plan. In particular, she asked for more control of the visitors and the *layas* they contract with to conduct offerings. She explained to me that it is not only the trash that the visitors leave from their picnics (plastic bottles, cups, plates, cutlery, etc.) but also the offering objects, such as glass bottles and candles, that they abandon that contaminate the mountain. To control this threat, the woman contended, the regional government must survey the offerings, and "to make this happen we need to put pressure on the authorities." The tensions caused by the offering rituals' environmental impact are amplified by the participants' perception of climate change, which many interpret as a local rather than global phenomenon (Paerregaard 2018a). While attributing the rising temperatures to the regional mining industry and Huancayo's air pollution, they believe that Huaytapallana's glacier melt is caused by the visitors' energy, particularly when they engage in physical contact with the ice. Some even believe that humans' increasing presence on Huaytapallana eventually will lead to the mountain's death, as the pilgrim in the opening scene of the introduction claimed. From this perspective, the offering ceremony endangers Huaytapallana's prosperity as much as it contributes to it.

While attending the pilgrimage on June 21 three times between 2014 and 2017, I have had the opportunity to observe the regional government's efforts to control the visitors' access to the mountain area and, on the other hand, the shamans' and the pilgrims' attempts to adapt to the new measures. Listening to Pedro, who is the pilgrimage's main actor and is recognized as a *laya mayor* (Quechua: principal ritual specialist), helps me understand how Huancayo's shamans perceive such issues as climate change, environmental contamination, and the regional government's efforts to regulate the ritual activities he directs. During the interview Pedro told me that he always had a creative mind and that his parents sent him to Huancayo's art school (Bellas Artes) when he was young. However, rather than stimulating his curiosity about modern Western art, the school opened his eyes to Andean art, in particular *la cosmovisión andina* (see chapter 1).

Even though Pedro's parents were both ritual specialists, he believes it is his personal capacity to communicate with the nonhuman world that qualifies him as *laya* mayor. In 2014 Pedro said to me: "You anthropologists want to understand the Andean world. I feel and sense it." He continued: "It is the feeling that you connect with the *apu* and that it responds which proves you are able to establish dialogue." Furthermore, *layas* must be capable of communicating with *pachamama* and other natural forces and feeling the magnetic vibrations of Earth, a capacity that enables them to mitigate the impact these have on humans in the form of stress. "A *laya* must be able to transmit this feeling to other people and in this way motivate them," Pedro told me. He also emphasized that the Andean cosmovision is very different from the Western worldview. "To become *laya* I had to decolonize my way of thinking. You must give up the idea of controlling the world. It is chaos. Instead of trying to change it we must live in the present," he explained. Pedro's interest in Andean cosmology has also inspired him to revisit his notion of counting and to turn on its head the Western tradition of privileging even numbers. "It's not the even but the uneven numbers that matter," he pointed out to me. Finally, to Pedro decolonizing his way of thinking implies revising the Western conception of global warming. Asked what he thinks about climate change and its impact on Huaytapallana, he replied, "Climate change is a political invention. The Earth has its own evolution. If Huancayo runs out of water people will have to move or even die." Life goes on, Pedro asserted, even if Huaytapallana dies.[14]

Pedro has been a practicing *laya* since the mid-1980s, when he established Museo Wali Wasi (Quechua: House of Spirits), a ritual center located in his natal neighborhood of Umuto on the outskirts of Huancayo. The museum contains an altar to worship Andean deities and a collection of Pedro's art pieces, and it figures in several tourist guides on Huancayo. It serves as a space for Pedro to meet his local followers, demonstrate his ritual talent, and extend his international network. "I have received people from all over the world and from all religions. From Tibet, Thailand, the Middle East, Israel, and North America. Once a couple of nuns from Huancayo even came to see me," he told me. Peru's politicians have also shown an interest in Pedro. Some years ago the country's first lady asked him to organize an offering ceremony. "I received an invitation from Eliana Karp [the wife of former president Alejandro Toledo] to make an offering at Pachacámac," Pedro recounted.[15] He described his vocation as *laya*, refusing labels such as *sacerdote andino* (Andean priest) that are often used to denote ritual specialists in the Peruvian Andes. "I'm *laya*, not shaman," Pedro pointed out, using the Huanka rather than the Western term for ritual specialist. It is important to make this distinction, he contended, as some people regard him as an associate of evil forces. "Some years ago, the church launched a campaign against me. They called me the devil. But I don't mind. I take it as recognition of what I do," Pedro said, declaring: "I don't have issues with other religions."

A decade ago there were few *layas* in Huancayo, but today there are more than two hundred, among whom approximately 10 percent are women. Most of them speak both Spanish and Quechua, though not all of them are fluent in the latter. Moreover, some *layas* come from other regions of the Peruvian highlands and are unfamiliar with the Huanka dialect. Nevertheless, Pedro views the new *layas* as competitors. He said: "There is competition, and you must watch out. Once I had to ask someone who pretended to be *laya* to leave. I had called for a ceremony and there were many people here [in Wali Wasi]. Then the man started to distribute his business card. Imagine, competing here in my place." As initiator and promoter of the June 21 offering ceremony at Huaytapallana, however, Pedro has carved out his own niche in Huancayo's market for ritual specialists, an effort that he proudly recalls.[16] "In 1987 I invited people to celebrate Andean New Year here in Wali Wasi, but they stayed all night and left it in a mess. The following year we did it somewhere else in Huancayo and in 1994 we started to organize it at Huaytapallana," Pedro told me, adding: "The first year we were less than ten persons. Now we are almost one thousand." Although Pedro hopes that his pioneering role in the offering ceremony will ensure his future legacy as *laya*, he recognizes that the growing participation is a source of concern. In particular, he finds it troubling that many visitors regard the offering as a payment they make in exchange for a specific favor. Rather than calling the offering *pagapu* (see chapter 1), which is common in the Andes, Pedro prefers the Quechua term *unkapaq*, meaning "to give." "You give the *apu* something to express your gratitude, not to ask for a favor," Pedro insisted. When asked what might be done to change this misconception of offerings, he shrugged his shoulders and replied: "Nothing. That's the way people are."

I also interviewed Carlos, the anthropologist I introduced in the opening scene, who teaches anthropology at Huancayo's major university (Universidad del Centro del Perú). Carlos has emphasized his identity as Huanka since I first came to know him in the early 1980s, and in the interview, he revealed that for more than a decade he had been Pedro's apprentice, attending the ceremony at Huaytapallana every year. Even though Carlos has learned a lot, he still has a long way to go before he will become *laya*. He told me that *layas* are recognizable by their discourse, their capacity to move people, and the effects of their offerings. "Only I can tell when I'm ready. That's what Pedro has told me," he explained. To test Carlos, Pedro gives him exams, one of which is to take *ayahuasca* (Quechua: a brew of Amazonian plants that produces hallucinations). Some years ago he went to Piura in northern Peru, where a local shaman guided him through two *ayahuasca* séances. Carlos reported: "The first time I almost didn't return but I learned a lot about myself." To prepare for the second exam, which is to ascend Huaytapallana on foot alone at night, a trip that takes five to six hours, Carlos walked to the mountain's glacier with a group of students two years ago. He described the trip with excitement: "After a couple of hours I asked the students to go ahead. While walking alone I felt I was flying, like a pheasant." Carlos believes the experience is related to the fact that he descends from a family named Condor, and that he has the blood of a condor in his veins. He said: "Very soon I'll make the trip to Huaytapallana on my own. I've already spent the night up there several times. It's incredible."

Even though the art of conducting offering ceremonies and teaching anthropology are two very different challenges, Carlos views them as complementary. Unlike other anthropology professors who lecture students on Peru's social and political problems, Carlos encourages them to explore their cultural past. In particular, he instructs them in the Andean cosmovision and teaches them how to show Huaytapallana, *pachamama*, and other nonhuman agents respect. Several times he has even invited students to the offering ceremonies that he organizes at Ancalaya, a sacred place outside Huancayo. "Sometimes more than a hundred students come to Ancalaya. They are very interested in Andean cosmovision," Carlos explained. He has also helped students organize a movement called Movimiento el Apu (the Apu movement), which was formed in 2008 and won the elections to the university's student board in 2010. As a university professor, Carlos thinks of himself as a mediator between the students and the Andean world. "I'm merely a bridge," Carlos told me.

Pedro's and other *layas'* role in the growing contamination of the mountain's environment is especially problematic, as they find themselves competing on Huancayo's regional market for ritual specialists and as the regional government is trying to hold them accountable for the visitors' activities within Huaytapallana's protected area. Moreover, the utilitarian use of offerings for economic and political purposes has led to a commercialization of the ceremony and paved the way for a new generation of *layas* who act as entrepreneurial brokers rather than spiritual leaders. Finally, climate change compromises the *layas'* task as stewards of the human/nonhuman relationships by foregrounding the material rather than symbolic meaning of the gift and highlighting an unresolved problem in the Andean cosmology they advocate: the offering's anthropogenic effect. The environmental impact of the growing number of visitors generates a divide between, on the one hand, established *layas* such as Pedro, who introduced the offering tradition at Huaytapallana thirty years ago and who makes an ecumenical reading of the Andean cosmovision, and on the other, the many new *layas* who use it as a vehicle for economic, political, and personal gain and who cater to the commercial market for ceremonies. By defining the offering as a gift rather than a payment, Pedro underscores his own authority as *laya mayor* and distances himself from this development. However, by arguing that glacier melt is part of nature's evolution and rejecting the global discourse of climate change, he not only disapproves of the new *layas'* secularization of the offering ceremony but also positions himself as a climate change denier and leaves open the door for other *layas* who subscribe to a politically charged interpretation of climate change that blames the extractive industry and capitalism for Huaytapallana's sufferings. More importantly, by refusing to recognize Huancayo's environmental problems and the everyday concerns that nurse his followers' faith in Huaytapallana, Pedro neglects the meaning they ascribe to the mountain's ice as symbol of authenticity and purity (Paerregaard 2018b).[17]

On my return trips to Huaytapallana in 2015 and 2017 I observed a shift in both the regional government's regulatory interventions and the pilgrims' readiness to adapt to them. In 2015 access to the mountain area was still unregulated, and pilgrims could reach Lazuntay, where the main event took place. By the same token, no one interfered when people strolled onto the glacier or took a swim in the lake, and even though some of the shamans encouraged the participants not to leave trash on the ground ("otherwise the authorities point fingers at us," one shaman warned), their appeal had little effect on the pilgrims' behavior. In 2017, however, I noticed several changes. As in previous years the regional government had put up a control post in Acopalca, the last settlement before the entrance to the mountain area, but this time all cars and travelers were submitted to an exhaustive scrutiny by the new director of environmental protection and her assistants. The inspection came as a surprise to the pilgrims in several cars, who loudly contested the director's order to withhold the beers and other items they had brought because they could damage the environment.[18] The director told me: "We don't allow people to bring things like bottles and candles. When they come back, we return the items to them." She went on: "Of course, later we must clean up the area. Last year we picked up more than six tons of garbage." Moreover, because the regional government has restricted the access to Lazuntay, the pilgrims had to conduct the main event at a distance from the glacier, which could be viewed but not touched. Finally, and perhaps most important, several pilgrims explained to me that when making their offerings to the mountain, they only leave organic items on the ground. One woman told me that she unpacks her offering gifts, leaving only food and sweets while keeping the plastic package, which she takes with her when she goes home. She said: "If we take care of the mountain, it will remember and reward us." Arguably, the adaptive capacity of transactional metabolism is paying off.

TRANSCENDING THE SACRED

The many economic interests, political viewpoints, and cultural perspectives that are at stake in the offering ceremony on Huaytapallana reveal the uncertainties that climate change and glacier melt are creating in Huancayo. In particular, they illuminate the confusion climate change causes in people's perceptions of the city's newly acquired prosperity and the role the participants in the offering ceremony play in its deteriorating environment. As in other parts of the world, many link climate change to change at the local (or regional) rather than global level and attribute its cause to their own agency or to nature's own process of change. At the heart of this confusion, and more specifically of my travel companions' responses to the contamination of Huaytapallana's environment and their reflections on Huancayo's future prospects, is a revision of the meaning of the sacred separable from the human realm and the idea of religion as a discrete human activity unaffected by other social and cultural processes. Paradoxically, while climate change challenges conventional forms of belief, it also opens new doors for religious practices and imaginaries that are driven by people's concern for the environmental change they are experiencing and the economic and political opportunities that emerge in the wake of this change.

In this brave new anthropogenic world, mountains are still conceived as forces with the capacity to influence the lives of other beings, but rather than seeking their protection or fearing their anger, Andean pilgrims now regard them as victims of human activities and worry about their survival. In a similar vein, the offering ceremony has changed meaning. While it previously aimed to appease the mountains and ensure the reproduction of human life, it now serves as an act to regret the harm humans are inflicting on the rest of the world. As a result, the offering ceremony becomes an act of self-reflection and repentance in which the giver not only pays tribute to the deities but also redefines their mutual relationship. Rather than constituting a mere symbolic act to communicate with the superior powers, as suggested by the classic theories, the offering has evolved into a setting wherein the giver shows concern for the taker's well-being and rethinks humans' role in the planet's future. By recognizing not only the footprints that the offering objects leave on the ground but also the consequences their physical remains have for the environment, the giver recasts the meaning of the sacred that becomes a referent to the shared effort humans and other beings make to save the planet from degradation and destruction.

FIVE

Quyllurit'i

THE GLACIER THAT SHINES LIKE A STAR

"Look, that's the South Star," Pablo says while we are looking southward at the sky in a frost-clear night. Then he turns eastward, pointing at two long lines of light that are making their way up to one of the nearby glaciers. "And that's the *ukukus* on their way up to the ice," Pablo continues. After a few minutes contemplating the spectacular view of humans dressed as bears walking with torches through the star-illuminated Andean landscape, he adds, "We're cold here. Imagine how it would be to spend the whole night up there on the ice." I nod and reply: "You're right. It must be an incredible experience but right now I'm glad I have a tent and a sleeping bag."

This conversation took place on May 28, 2015, at a remote pilgrimage shrine. Pablo and I had left the city of Cusco the day before in a well-packed minibus that took us on a four-hour ride to the town of Ocongate, where we spent the afternoon in the company of a group of bear-dressed men paying homage to an image of Quyllurit'i in a small chapel.[1] After the ceremony a group of local women offered us food and chicha, whereupon the bear-dressed men first danced to the music that a band was playing for a while. Then the dancers and musicians all climbed onto a truck that would take them to a small town called Mahuayani. The following day Pablo and I followed them in a *colectivo* with five other passengers who all were embarking on an unforgettable experience: the annual pilgrimage of Quyllurit'i.

Pablo is a Peruvian anthropologist from Cusco and an old friend of mine who had visited Quyllurit'i in the early 1980s. He was therefore familiar with the event, which is one of the largest pilgrimages in Peru. During the three days the gathering lasts pilgrims from neighboring villages as well as from the rest of Peru and even other countries in South America and Europe visit the shrine, which they reach after a three-hour walk from Mahuayani. Like

FIGURE 14. Pilgrim camp at Quyllurit'i at the foot of Qulque Punku, in the southeastern highlands of Peru, 2016. Photo by author.

Pablo and me, many sleep at Sinak'ara in tents they have brought with them or bought from traders in Mahuayani (see figure 14). Some defy the cold, spending the night outside even though the temperature drops below zero. However, whether inside or outside few get their beauty sleep, as the sound from the music bands, the dancers' whistling, and the fireworks people set off never stop. Yet to my surprise, we didn't see a single person drinking alcohol during the three days the pilgrimage lasted, although it is an integral part of religious festivities in many parts of Peru. Despite (or perhaps because of) the musicians' and dancers' persistent and relentless noise, this lent a feeling of not only solemnity and sobriety but also communality and vitality to the event which, according to many of the pilgrims I talked with, fueled them with energy to go back home and take up their daily chores.

"You don't see any police and authorities here. The *ukukus* [the bear men] keep order," Pablo says to me, adding, "Alcohol is not allowed. You're expelled if they catch you drinking." He explains: "The *ukukus* always carry a whip which they use to punish people who disobey. They whip your shins." Pablo's warning came true the next day. While I was taking photos of a group of dancing *ukukus* one of them whipped my right shin. The man literary taught me to behave as a good pilgrim.

I have attended pilgrimages in other places in Peru as well as in Europe but never witnessed such a spectacular scene as Quyllurit'i, which displays a mixture of religious asceticism, folkloric diversity, and physical bravery in a landscape where the boundary between humans and nonhumans blurs. The three previous chapters have described offering rituals as a metabolic replication of a water supply on which their participants depend either directly or indirectly. By contrast, in Quyllurit'i most pilgrims only establish symbolic contact with the mountain's glacier, which has no material importance for their livelihoods and daily lives. Yet to the *ukukus*, who engage physically with the glacier, this represents a living substance that connects them to the nonhuman world. Rather than paying tributes, taxes, and tariffs to smooth the water flow, as the villagers in Tapay and Cabanaconde do, or making offerings to ask for goods and favors, as the pilgrims of Huaytapallana do, the *ukukus* use their own bodies to metabolize the ice and its spiritual power, generating a transcendental experience that sometimes takes their lives. However, as in the previous cases, climate change, glacier melt, and environmental change disrupt the ritual's performance, which is shaped by a drama that happened more than two hundred years ago and involves a fabric of social and ethnic actors.

INDIANS AND MESTIZOS

Quyllurit'i (Quechua: Snow Star) is a pilgrimage comprising a set of ritual activities that unfold in different locations southeast of the city of Cusco during the three-week period between the movable feasts of the Ascension and Corpus Christi (Sallnow 1987). The main event takes place three days before Corpus Christ (celebrated on the Thursday after Trinity Sunday) at a site called Sinak'ara that contains the sanctuary of Señor of Quyllurit'i, which has given its name to the pilgrimage. At Sinak'ara there is also a grotto that houses the Virgin of Fatima whence the visitor can view the glacier of Mount Qulque Punku (Quechua: Silver Gate) to the north and to the south the snowcaps of Mount Ausungate, the region's tallest and most renowned mountain.

No written sources exist that can document the exact origin of the pilgrimage and its legends (Salas Carreño 2006). However, some scholars claim that they can be traced back to Túpac Amaru II, an indigenous uprising that was named after an Inca ruler and that took place between 1780 and 1783 (Brachetti 2002). The movement fought to reestablish the Inca empire but was crushed by the Spaniards, who executed its leaders and many of their followers.[2] Michael Sallnow writes: "The revolt was in large measure a conflict between Indians, mobilized for one side or the other by kurakas [indigenous leaders]. Partly as consequences of this, the geographical distribution of the rebellion was highly uneven" (1987, 214). The uprising's native divide was reflected in the original organization of Quyllurit'i, which up to the 1990s was enacted by two opposing groups: Quispicanchis, a predominantly pastoral community that supported the rebels in 1780, and Paucartambo, an agricultural community that offered them fierce resistance (1987, 214). It is likely that in order to uproot the local population's belief in mountains, it was the Catholic church that created the story about Jesus's revelation at Sinak'ara at the foot of Ausangate, one of the Inca's most sacred mountains. According to the legend, the Savior appeared first disguised as a white child in front of an Indian boy who was herding animals and later as an image that is known as Señor de Quyllurit'i (as the image is called in Spanish) or Taytacha Quyllurit'i (as it is called in Quechua). Yet rather than converting Andean people into devoted Catholics, the legend merely added a Christian layer on top of the indigenous religion that still prevails in Quyllurit'i even though it is recognized as a Catholic saint (Poole 1990).

Following this interpretation of the pilgrimage's origin, Quyllurit'i is the product of the religious syncretism that emerged from the Spaniards' attempts to suppress the adoration of Andean deities and control the native population. During the colonial period the Catholic church recurrently proclaimed the revelation of the Savior in front of members of Peru's ethnic minorities in the cities after natural disasters, as happened in 1655 when an earthquake hit Lima and gave rise to the legend of Señor de Milagros, or in remote mountain areas after important political events such as Túpac Amaru II, which may have produced Quyllurit'i.[3] Unlike the legends of many of the miracles that propel today's predominantly Christian processions in Peru's cities, it is the myth of the latter that fuels many of the country's contemporary pilgrimages to Andean shrines such as Sinak'ara that blend the preconquest and Iberian heritage. Commenting on the religious nature of Quyllurit'i, Sallnow writes: "These two facets of miraculous shrines, Christian and Andean, are locked in a perpetual embrace. Christian icons become Andean, entrenched in the symbolic landscape of the peasantry and imbued with vernacular meanings and significance, yet it is precisely this deep-rooted and pervasive symbolism that makes them such valuable trophies for religious and social élites. There can be no ultimate compromise or accommodation here: only a persistent split in religious belief and practice, a manifestation of the relentless cultural antagonism that history has visited upon the Andean peoples" (1987, 269).

The religious syncretism that shapes the legend of Quyllurit'i comes to the fore in competing versions that have been fabricated by the church, the local villagers, and the many pilgrims. The official version stresses that the light shining at Sinak'ara briefly turned into a crucified Christ over a tree called *tayanka*, which had the form of a cross. But as this was taken to Spain upon the demand of the Spanish king and thereafter never returned to Peru, a movable image of Señor de Quyllurit'i was made, which now is housed in the parish church of nearby Ocongate. To complicate the matter further, a replica of the image called Señor de Tayankani is kept at a place called Tayankani, named after the tree where the Savior first appeared. A final point of reference in the pilgrimage's Catholic universe is the chapel of Mawayani, the point of the pilgrims' disembarkation for the sanctuary (Ceruti 2007; Sallnow 1987). Local interpretations, however, present the legend somewhat differently. Rather than pointing to a tree that turned into a cross, they relate that the Savior revealed himself on a crag at Sinak'ara. In the first half of the twentieth century the church tried to frame the movable image in Ocongate as the miraculous center of the pilgrimage (Salas Carreño 2014, 202), but in the religious mind of most pilgrims the crag on which the Savior appeared is still the pilgrimage's epicenter. Located behind the altarpiece of the sanctuary's main chapel, it has a painting of a crucified Christ, which is the focus point of most pilgrims and which Quechua speakers address as *taytacha* (Quechua: little father) rather than *señor* (2014, 190).[4]

The distinct title pilgrims use to address the image is a reminder of the different Christian and Andean histories that Quyllurit'i represents. In many of the pilgrimage's activities the boundary between the two religious (Andean vs. Catholic) and social (Indian vs. mestizo) worlds is blurred, but the events' multiple ritual layers are nonetheless an indicator of a latent tension between the official and the native readings of the image's legend and the meaning of Quyllurit'i. As Sallnow puts it: "What gives the Qoyllur Rit'i fiesta its distinctiveness is the second level of ritual activity: a complex sequence of rites, processions, and minipilgrimages focused on the Sinakara shrine and embracing all its associated sacred sites and religious images. The second set of ritual activities is superimposed on the first, though no means all the visiting pilgrims participate in it in its entirety" (1987, 224). As I discuss later, the massive popularity that Quyllurit'i has gained in recent years is intricately related to the subtle ways that Andean cosmology and symbolism are ritualized at the sanctuary and that Qulque Punku's glacier is associated with ideas of sacredness. In this more modern version, it is not God's luminosity but the sparkling light of Qulque Punku's ice that shines on the sanctuary and the pilgrims. From this perspective it is the glacier's light and its role as a stage for the pilgrims' contact with the nonhuman world that has given the name to the pilgrimage and that animates its divinity.

Quyllurit'i's competing theological interpretations are also reflected in its geographical organization. According to Sallnow, Quyllurit'i is partitioned into two sectors: "the valley [Sinak'ara] with its miraculous, Christianized crag, and the encircling glaciated peaks, northern outposts of the Ausankati range" (1987, 238). He goes on: "To understand why Qoyllur Rit'i, of all regional shrines in Cusco, has become the focus for such an elaborate cultural drama, it is necessary to unpack the symbolism of this sacred core" (1987, 238). One way of doing this is to interrogate the social and cultural meaning of the *ukukus* and the many dancing groups' costumes and choreography, which reflect not only the pilgrimage's religious syncretism but also the wider sociocultural context it taps into. Until the 1970s most pilgrims were from the surrounding rural Quechua-speaking communities that came in groups, each carrying a small image of Señor de Quyllurit'i escorted by dancers (Sallnow 1974). According to some scholars, the dominating groups were identical with the two opposing ethnic communities that took part in the Túpac Amaru uprising during Spanish colonial rule: Paucartambo and Quispicanchis. Framed as *naciones* (Spanish: nations), they represented a classical Andean dual division in competing moieties (Brachetti 2002).[5] In the 1980s a third *nación* (Canchis) was added, and today there are a total of eight *naciones* named after their regional origin: Paucartambo, Quispicanchi, Acomayo, Paruro, Canchis, Urubamba, Anta, and Tahuantinsuyo (from the city of Cusco), whose members come from a variety of places (Salas Carreños 2020). Organized as grassroots associations, the *naciones*, which are tied to particular regions or urban neighborhoods in Southern Peru where they recruit their members, constitute the structuring body of the pilgrimage. Yet even though they are united in an organization called Consejo de Naciones. Peregrinos al Sanctuario de Quyllur Rit'i - Sinak'ara formed in 2004, and even though it sometimes coordinates the activities of the eight *naciones* to defend the interests of the pilgrimage—as last happened in January 2016, when they called for a march in the city of Cusco to protest against mining in the Sinak'ara area (see later)—these act as independent units.

Together the *naciones* mobilize more than eight thousand musicians and dancers grouped in *comparsas* and recognizable by their distinct way of dressing and dancing (Mendoza 2010). As described earlier, others act as *ukukus* (also called *pablitos* or *pauluchos*), young men dressed as human/ bear hybrids who use whips to enforce the pilgrimage's unwritten rules, including a ban on drinking alcohol. While the *ukukus* play an ambivalent role as both rule enforcers and tricksters in the pilgrimage, the *comparsas'* costumes and performance display a burlesque image of some of the most notorious social actors in Peru's republican history (Ceruti 2007).[6] Exploring how this scenery unfolds, Poole writes: "The interplay between this passively observing—and often uncomprehending—Western gaze, and the active Andean voice of dance, forms a crucial part of the script through which Qoyllur Rit'i, and particularly the dances performed at Qoyllur Rit'i, constructs its special sense of hierarchy and social identity for the pilgrims who attend. Rather than representing a simple 'confrontation' between 'two' cultures—the conquered and the conquering, the Western and the Andean—this dialog conveys a critical sense of unconscious complicity or collusion between their respective modes of interpreting social, hierarchical organization" (1990, 99).[7]

Syncretism also pervades the pilgrimage's organization, which aside from Consejo de Naciones comprises a Catholic brotherhood called La Hermandad del Sanctuario del Señor de Qoyllur Rit'i, which is more than eighty years old, takes care of the sanctuary, and has a strong say in the pilgrimage's management (Brachetti 2002). Guillermo Salas Carreño writes: "Founded in 1940 and recognized by the Catholic Church, the Brotherhood controls access to the shrine's church, and until recently had authority over all aspects of the pilgrimage" (2014, 199). Its foundation led to a restoration of the image, which was blessed by the archbishop of Cusco in 1944, whereafter the number of pilgrims began to swell (Sallnow 1987, 215).

The presence of numerous businesspeople adds a third dimension to the pilgrimage's cultural and religious hybridity. When the pilgrims arrive at Mawayani, a small town located at 4,000 meters where the pilgrimage commences, they are met by people selling food and equipment to camp out at Sinak'ara and renting horses to carry their luggage or be ridden to the shrine. Like several of its neighboring towns, Mawayani has a large community of Evangelists who view the adoration of not only Catholic images but also

Andean deities as sinful and therefore consider the pilgrimage an idolatrous practice, but nonetheless see it as an opportunity to make an income. According to Salas Carreño, "The pilgrimage has turned out to be a business opportunity for the people of the town of Ocongate and the surrounding communities, who have installed restaurants, sell souvenirs and cloths, and provide other services. In doing so, the people of Ocongate moved their pilgrimage to another time of the year. Others, mainly people of the surrounding rural communities who rent their horses or sell food, are Evangelicals. While they regard the pilgrimage as an idolatrous practice, they do not miss the monetary opportunity to gain income" (2014, 189).

After four hours' walk, the pilgrims arrive at Sinak'ara, where another group of businesspeople are waiting. Coming from Puno and other places around Lake Titicaca and of Aymara origin, they sell *alasitas* (see chapter 1 and 4), including miniatures of houses, cars, and other goods, as well as copies of forged professional titles, which the pilgrims desire and hope to acquire or obtain in the year that follows (Stensrud 2010). Farther up another group of Aymara-speaking businesswomen make money on the pilgrims' wishful thinking by letting them use stones to design the interior of their future dwellings on the ground. A few years ago, the brotherhood that manages the shrine where the image of Quyllurit'i is kept built a small house at Sinak'ara called Banco del Señor de Qoyllur Rit'i, which it rents to fortune tellers and businesspeople acting as bankers. Selling counterfeit foreign banknotes, these financial entrepreneurs offer phony loans in the name of Quyllurit'i to pilgrims who hope they will bring them luck. In the eyes of many pilgrims such commercial activities defy the religious ideals of Quyllurit'i, but others engage in them precisely because they have faith in the image's divine power and capacity to create fortune and well-being.

Little is known of how the legend of Quyllurit'i impacted the native and the mestizo populations after its claimed conception in 1783, which may be an indicator that the image received little attention outside the sanctuary's close surroundings. But in 1935 the fading image on the crag at Sinak'ara was retouched, and in the second half of the twentieth century the pilgrimage has slowly but steadily gained momentum (Sallnow 1987, 214). Until three decades ago the pilgrims mainly came from the neighboring communities and the rest of the Cusco region, but in the 1990s the event has attracted a steadily growing number of people from other parts of Peru, and today it is one of the most attended pilgrimages in Peru with almost one hundred thousand participants.

It is impossible to unpack Quyllurit'i's syncretic texture and separate its Catholic from its Andean dimensions. Even so, by situating the Savior's revelation and thus the legend's ground zero at the back of one of the most powerful *apus* in the Andes, the Catholic church tapped into the religious universe of the indigenous population and linked Christ's crucifixion and resurrection to Andean notions of sacredness and power. Thus, María Constanza Ceruti links the pilgrimage's location to the mountain of Ausungate, one of the highest Andean peaks that plays an important role in Inca mythology. Ceruti writes: "One of the main reasons why the place is recognized as sacred is that it is perceived to be under the sphere of influence of nearby Ausangate, a mountain most sacred to the ancient Incas and to the modern Quechua ritual experts, invoked during their initiation ceremonies and divination rituals" (2013, 4). The suggestion that the pilgrimage taps into its divinity from Ausungate and the mountain's capacity to produce water resonates with other studies of Quyllurit'i. David Gow contends that when people pay respects to the former, they also do it to the latter. In his study of the pilgrimage, he observed: "Close to the sanctuary, next to a lake, people continue to make despachos [offerings] to Apu Ausangate" (1974, 57). He goes on: "When one is praying at the Qoyllur Rit'i sanctuary, one is looking at Ausungate at the same time. The word rit'i (snow) refers to both: The Lord and Apu Ausangate" (1974, 57). According to these scholars, then, Ausungate is synonymous with water and ice and the power to generate life, and so is Quyllurit'i.

Hence, even though Quyllurit'i was crafted as a reminder of Christ's suffering and God's almighty power, it is intimately associated with *apus* and their religious and political power. Sallnow goes as far as to argue that "given the importance of Mount Ausankati in Andean cosmology from pre-Hispanic times to the present, it is perhaps surprising that the Christian shrine of Qoyllur Rit'i was not founded until 1783, 250 years after the Spanish conquest" (1987, 213). He finds support for this claim in the interpretation of the pilgrimage by Qamawara, a neighboring community that views the shrine's divinity as an entirely Andean affair and that explicitly locates it at Ausungate. Exploring how this local indigenous reading of Quyllurit'i challenges its official version, Sallnow writes: "Finally, in keeping with this stress on the autochthonous origins of the cult, the Qamawaran account established an overt connection between the taytacha and Apu Ausankati:

a pilgrimage to the Christian shrine is deemed to be at the same time a visit to the pagan mountain deity" (1987, 211). Epitomizing Ausungate's ice and snow, the divinity of Quyllurit'i is thus a reconfiguration of the *apu's* spiritual energy.

Pilgrims contribute to the ritual vivification of ice and meltwater even before they reach the sanctuary. On their ascent from Mawayani to Sinak'ara, many stop at the water of the creek that runs downhill from Qulque Punku's glacier. Driven by an idea that this can heal, they hope to acquire some of the creek's properties by bathing or washing their clothes in its meltwater. Ceruti writes: "The majority of the women start to wash the hair—in many cases with shampoo—at the same time they rinse the faces and hands of their children. The men often take off the shirt and wash the torso, and the head too. Some pilgrims—including old women—prefer to bath completely naked despite the intense cold, moved by the faith not only in 'the Lord's water does not make you sick' but also in the curative powers it possesses" (2007, 20). Other pilgrims stop at a spring called Agua del Señor on the way to Sinak'ara to drink its water, which they believe originates from Qulque Punku or Ausungate. Angela Brachetti reports that the water is said to have spiritual power and that "it is famous for being miraculous, just as the ice of Ausangate's glacier" (2002, 90).

The association of Quyllurit'i with the metaphysical qualities of mountains and their ice and the water they produce comes to the foreground in the mythical representation of *ukukus*. As described previously, these young males who are named bears but whose faces and heads are covered with alpaca wool and who carry masks that originally were made of alpaca or llama wool are responsible for keeping the discipline within their nation's dance troupe and the pilgrimage by and large. Paradoxically, they also act subversively and take on burlesque characters. The *ukukus'* ambivalent role is underscored by their image as liminal beings who simultaneously are humans and beasts and who acquire the force of superhuman agents by interacting with the glacier and the ice (Salas Carreño 2014, 206).[8] In Ceruti's interpretation the *ukukus'* performance on the glacier even adds power to its divinity: "The sacredness of the ice is enhanced with the healing and fertilizing qualities attributed by the reverent footsteps of the brave native bear-men" (2013, 8). Whether the *ukukus* are believed to enhance the glacier's spiritual capacity or are simply viewed as mediators between the human and the nonhuman worlds, their engagement with the ice puts them in direct contact with the power that controls the hydrological cycle and produces water.

On the glacier the *ukukus* establish rapport with another dubious figure: the *condenados* (Spanish: condemned), that is, dead humans whose souls never have made it to the other world because of their previous sins and who therefore restlessly wander on the *puna* in search of redemption. Common to both is the attempt to gain the *apus'* goodwill, the *condenados* to seek redemption and the *ukukus* to access the sacred, which in a certain way makes them soulmates. Sallnow writes: "The ukuku dancers at Qoyllur Rit'i are likewise intimately associated with condenados. It is widely believed that the glaciers in the vicinity of Mount Ausangaki are infested with condenados, for by climbing the mountain, naked and in chains, to reach the silver cross near the summit, they can obtain forgiveness from the apu and release from their tortured state of living death" (1987, 219–220). Ceruti goes as far as comparing the *ukukus'* almost inhuman undertaking on the glacier with that of the *condenados*, suggesting that "their impressive demonstration of physical endurance is meant to be in itself an offering to the nearby mountain spirits or *Apus*, especially to *Apu Ausangate*" (2013, 5). Brachetti also finds proof that when entering the glacier, the *ukukus* are in fact reaching out for Ausangate. She writes: "Facing Ausungate they [the *ukukus*] ask for the absolution" (Brachetti 2002, 98).

The *ukukus'* mediation of the human-nonhuman worlds crystallizes on the pilgrimage's main day when the bear men first place the cross on the glacier and later return it to the sanctuary. Sallnow describes this crucial part of the pilgrimage in the following way: "The snow-covered slopes of Ausankati, as noted, are peopled with condenados vainly attempting to scale the peak and attain salvation. The taytacha, too, ascends the glacier in the shape of a cross—or, nowadays, crosses—but he is reclaimed and returned to the sanctuary by the ukukus, themselves condenados and hence the only beings capable of performing such a feat. Imprisoned in his rock abode just beneath the peaks, permanently denied the possibility of release from his condition, his living corpse radiates a power that can be tapped by human beings to alter their private or collective destinies" (1987, 238). Still, unlike *condenados*, *ukukus* belong to the human world, and their courageous endeavors to engage with the former's tortured souls can also be viewed as an effort to protect the pilgrims from the danger they are exposed to when gathering at Sinak'ara and to safeguard the Savior's image from the threats the unknown domain of Andean nonhuman agents.

Arguably, the *ukukus'* ritual performance is the key to understanding the symbolic interlocking of the glacier and the crag and Quyllurit'i's syncretic nature. As Salas Carreño points out: "Even though the rock and the glacier can be seen as different places, there are several ways in which they are related to each other. The first one is the location of the rock. It is so close to the glacier that the latter's presence is hard to ignore while in the shrine. The second relation is given by the very name Shining Snow, which also points to the glacier. These relations make the rock/image of Christ an index of the glacier, that is, the rock points to the unavoidable presence of the glacier" (2014, 206). Or, to phrase it differently, by linking the glacier to the shrine, not only do the *ukukus* connect Andean ideas of divinity and spiritual power to Christian axioms of redemption and salvation; their ritual interaction with the glacier also vivifies the meltwater that this produces.

METAPHORICAL METABOLISM

But the *ukukus'* transgressive performance on Qulque Punku's glacier when they first carry and place their crosses on the ice and later return them to the chapel at the sanctuary not only vivifies water; it also ritualizes water metabolism, though in a different manner than described in the previous chapters. Rather than metabolizing material objects to obtain goods such as water, as the villagers of Tapay and Cabanaconde do, or personal services and favors, as the pilgrims of Huaytapallana do, the *ukuku*s seek spiritual redemption by engaging in a symbolic "change of matter" of their own bodies. Some *ukukus* even end up as offering objects themselves. While the goal of reciprocal and contractual metabolism is to speed up the water flow, and while transactional metabolism uses this as a medium for individual gain, the metabolization of ice is the very process whereby *ukukus* achieve absolution in metaphorical metabolism.

In a collective catharsis and the culmination of the pilgrims' search for forgiveness, the *ukukus* ascend the glacier and place Christianity's most powerful symbol in its glittering ice, which is perhaps the pilgrimage's most evident effort to unite the Christian and Andean worlds (Brachetti 2002, 92) (see figure 15). The scenery is truly magic. From Sinak'ara Pablo and I—along with thousands of other pilgrims—watched how the eight nations of bear-dressed men in single file made their way to the glacier carrying torches and crosses. Halfway up the mountain slope, the *ukukus* split to reach the glacier's three tongues. Several years ago, the *naciones* agreed to divide these so that (viewed downhill from the top of the glacier) three groups of *ukukus* (from the

FIGURE 15. *Ukukus* returning from Qulque Punku's glacier, in the southeastern highlands of Peru, 2016. Photo by author.

nations of Paururo, Urubamba, and Acomayo) ascend its left tongue, three groups (from the nations of Quispicanchi, Anta, and Canchis) ascend the tongue in the middle, and two groups (from the nations of Tawantisuyo and Paucartambo) ascend its right tongue.[9]

Once they have climbed the glacier the *ukukus* set off fireworks to galvanize an atmosphere that is both solemn and exhilarating. They also engage in a series of ritual activities that have developed over the past five decades. In the 1960s the only *ukukus* represented at the pilgrimage came from the communities of Paucartambo and Quispicanchi, which fought ritual battles against each other on the ice. Juan Andrés Ramírez portrays the two groups as soldiers in warfare and the glacier as their battleground: "These two powerful armies of Ukukus, under order of their highest chiefs starts the ascend toward the snowcap, each troop on the path it corresponds" (1969, 85). Ramírez recounts that while those from Paucartambo went to "the snowcap's left side," those from Quispicanchi went to "its right side," which supposedly is viewed from below (that is, the opposite of my description earlier.) He reports that the ritual peaked when at 2:00 a.m. "the two sides meet and begin the famous guerilla throwing snowballs and reaching at point of grabbing each other body against body" (1969, 86). Importantly, the *ukukus* from Paucartambo and Quispicanchi not only ascended the same spot on the glacier (which at that time almost reached the sanctuary) but also shared the same cross, which they placed on the ice in a common effort. According to Ramírez, after an "arduous" fight the winning group would bring the cross from the ice down to the sanctuary as proof of their victory, while the rest of the *ukukus* would follow them carrying ice blocks on their shoulders as a symbol of penance.

In 1984, when Sallnow did fieldwork and when *ukukus* from Canchis also were present on the glacier, the rituals had changed. Observing the performance of the three *ukuku* groups, he reports: "Some were kneeling singly or in groups of two or three, facing the summit, lighted candles planted before them in the snow, their lips moving in silent prayer. Others engaged in horseplay, throwing snowballs at one another and sliding down snow chutes with yells and shouts" (Sallnow 1987, 213). Sallnow also writes: "An ukuku visiting the glacier for the first time was obliged to receive a *bautismo* 'baptism' in the form of three strokes of the whip on the rump administered by the ukuku captain of his nación. After such a whipping, the victim kissed the whip and embraced the whipper" (1987, 213). Two decades later Brachetti observed similar activity: "Each nation has a place reserved where they conduct some ceremonies such as baptizing new members or punishing somebody who has committed some form of sin in the behavior during the fiesta or during the year. Such actions are carried out by the group's corporal. Both the baptism and the punishment comprise three whips in the buttock with the sling" (2002, 98).

To endure the cold, the *ukukus* enter a state of trance in which they communicate with the ice and the metaphysical powers it contains (Shapero 2017). Recounting the experience of spending the night on the ice in clothing ill fitted to temperatures below zero, some *ukukus* told me that it requires physical endurance but that they view this as a challenge rather than an obstacle. They also said that their contact with the ice is a spiritual experience that fueled their lives with new energy. The sense of liminality and trance is enforced by the danger it implies about walking on the glacier at night, a risk that has been imminent since Ramírez made his study of the *ukukus'* ritual battles in 1968. He points out that "the fight is dangerous because, by neglect, they may die" (J. A. Ramírez 1969, 68). Commenting on the danger of falling into the glacier, Sallnow reports: "Each group followed a separate route to its snowfield. As they stepped off the moraine onto the glacier, the ukukus linked themselves into long chains with their whips to prevent their slipping backward on the treacherous, icy slopes" (1987, 213). Referring to the ritual battles the *ukukus* formerly engaged in, Sallnow writes: "Although deaths were not actively sought in this encounter, it was nevertheless apparently not uncommon for ukukus to be killed by falling into the crevasses as they chased each other across the glacier. Nowadays there is no battle" (1987, 228).

According to Ceruti, however, glacier accidents still occur, sometimes with fatal consequences, which she interprets as a (arguably unplanned) sacrifice to the glacier and the *apu*. Drawing on Flores Lizana's ethnographic account (1997), Ceruti writes: "The death during the ceremonial instant is oftentimes also interpreted as an offering demanded by the mountain spirits, which draws hopes of survival for the community. Even in the case of death of children due to high sickness or disappearance of ukukus in the glacier crevasses, physical death is conceived as an offering to the mountain spirits, cable of ensuring the abundance of rain and the fertility of the harvest" (2007, 31). Among pilgrims such accidents are sometimes explained as the glacier "eating" or "swallowing" the victims, whose fate is attributed to their lack of faith in Quylluri'ti. Several *ukukus* told Ceruti that "almost always, two or three pablitos [*ukukus*] are swallowed by the ice" (2007, 30). Another *ukuku* explained to Ceruti that when the ice recedes the bodies of some of the disappeared *pablitos* from previous years reappear. Imitating the rigid posture of a dead body, the *ukuku* said: "That's how you find them," adding that "these skinny guys, you notice that there ain't much to eat up there" (2007, 31), which, according to Ceruti, is a way of accounting for the *ukukus'* fatal destiny on the glacier as "not as a true death" but "the luck of existence that endures in the glacier's strangeness" (2007, 31).

At dawn the *ukukus* walk down to Sinak'ara carrying the cross of their *nación*, which they return to the chapel (see figure 16). Sallnow writes: "Around 8:00 A.M. the ukuku bands came down from the glaciers in formation, each with its cross, banner, and the Peruvian flag proudly held aloft" (1987, 228). He continues: "Each procession marched through the pilgrims' encampment to the cairn cross, then return to the sanctuary to deposit its cross" (1987, 228). Some *ukukus* also carry chunks of ice from the glacier, which they believe have the power to heal and bring luck and which they take with them to their communities. Previously some *ukukus* even brought chunks of ice to Cusco, where they gave it to the participants during the Corpus Christ festival celebrated on the city's main square in front of its cathedral two days after the pilgrimage, attesting to Quyllurit'i's linkage to Peru's colonial epoch and its syncretic history.[10] Brachetti relates, "It is a form of penance according to their belief and they say that the bigger the ice block the more serious are the sins" (2002, 98). Sallnow's account illustrates how the practice of cutting ice unfolds: "In the preparation for the descent, ukukus set about carving out chunks of ice from the glacier, sawing away laboriously with their rope whips; a few had brought

FIGURE 16. *Ukukus* celebrating their return from Qulque Punku's glacier, in the southeastern highlands of Peru, 2016. Photo by author.

picks for the purposes. The blocks of ice were then tied to their backs with their whips and carried down to the sanctuary as a penance. Many were simply dumped, but some people collected their meltwater in bottles and conserved it for medicinal use. A few ukukus even tied little blocks of ice to their dolls" (1987, 228).

Over the years other pilgrims have also started to ascend the glacier during the day. Ceruti reports from her 2007 study: "Pilgrims, including very young children, traditionally climb up the lower sections of the glaciers in daylight, to light candles on the ice" (2013, 7). She goes on: "Bare footed, kneeling on the snow, they may pass hours watching the candles burn. Their silent prayer is combined with a careful contemplation of the flame, intended to provide them with answers to their most intimate concerns" (2013, 7). The pilgrims celebrate their feat by engraving their names in the ice, which they then cut and bring with them upon their return to the sanctuary and later their homes. In this way, pilgrims and ice engage in a metabolic process that yields the former prosperity while turning the latter into water. The metaphorical circle of Quyllurit'i's metabolism has been closed.

Upon our arrival at Sinak'ara Pablo looked at the sanctuary and asked: "What happened to the ice?" Then he said: "When I was here in 1985 it reached almost down to the shrine. Now it's way up the mountain." Pablo's observation testifies to the impact of global warming on Qulque Punku's glacier and helps explain why Quyllurit'i has become an issue of concern for not only Consejo de Naciones and the brotherhood, which organize the pilgrimage and represent its interest, but also among the national and regional authorities, which see a growing need to regulate the movement and behavior of the many pilgrims on Sinak'ara to safeguard the glacier and protect the environment. Salas Carreño (2021) reports that glacier retreat has been a concern among Quyllurit'i's followers since the 1990s, while Ceruti noticed the melting of Qulque Punku's glacier as early as in the mid-2000s, when she reported that the ice was eroding and impeding the ascent to its higher part without professional equipment. Ironically, Ceruti also observed that diminished snow precipitation had reduced the risk of *ukukus* falling into the glacier's crevasses, as these were more visible than in previous years (2007, 30).

Still, climate change poses a serious threat to Quyllurit'i and in particular the ritual activities of *ukukus* (as well as ordinary pilgrims) on the glacier. Ceruti goes as far as claiming that "the ancestral rituals are changing due to the retreating glaciers caused by global warming, and Quechua people are in fear that once the ice is gone, the Lord of the Star of Snow will no longer listen to their prayers" (2013, 7). She concludes: "Global warming is a somewhat more silent destroyer of the heritage of the Andes" (2013, 3). But it is not only climate change that jeopardizes Quyllurit'i's cultural heritage. The environmental impact of the growing number of pilgrims also represents a major challenge. In the entire Sinak'ara area there is only one lavatory, with fewer than ten seats available for the huge crowd. Moreover, the lavatory sits on the top of the creek that transports the glacier's meltwater and that the pilgrims use downstream to bath and wash. The pilgrims' trash is also a problem. The brotherhood has put a sign at the entrance to the sanctuary asking people to protect the environment. Other organizations take care of cleaning the area after people have left.[11] Even so, the gathering's footprint is notable. Like Pablo and me, thousands of pilgrims camp overnight in an area around the sanctuary, and businesspeople put up tents where they cook and sell food, drinks, and other basic items. The number of traders from the Puno area selling *alasitas* has also grown, just as there are more "fortune sellers" in Banco del Señor de Qoyllur Rit'i.

The pilgrimage's climate adaptation has been complicated by UNESCO's declaration of Quyllurit'i as world intangible cultural heritage in 2011. The eight nations and the brotherhood have welcomed the initiative, which they see as evidence of the pilgrimage's importance in not only Peru but also the rest of the world. Nonetheless, given that global warming is undermining the pilgrimage's ritual practices by causing glacier retreat and that its popularity and attraction for new pilgrims is contributing to the sanctuary's environmental decay as well as the ice melt, the declaration raises a difficult question: How can a cultural tradition be preserved that is threatened by climate change and its own activities at one and the same time? The task of solving this conundrum has been entrusted to the Ministry of Culture, which is the institutional steward of Peru's cultural heritage and therefore also the official caretaker of Quyllurit'i. To accomplish its mandate and engage the pilgrimage's stakeholders in its work, the Dirección Desconcentrada de Cultura del Cusco (the ministry's local branch) occasionally invites the eight *naciones*, the brotherhood, and the Catholic church to meetings at its regional office in Cusco, where they discuss the organizational and logistical challenges the growing number of pilgrims creates and the environmental problems their activities on the glacier cause. The outcome of the meetings is a working plan (Plan de Salvaguardia del Patrimonio Cultural de Qoyllur Rit'i) that aims to preserve the cultural heritage of Quyllurit'i by improving the pilgrimage site's toilet facilities, sewage infrastructure, and garbage service and regulating the growing number of trade and business activities, especially the sale of *alasitas* and other "fortune" products.

Another, more sensitive measure concerns the restriction of the pilgrims' access to and activities on Qulque Punku, including a ban on cutting ice chunks from and burning candles and fireworks on its glacier.[12] Three of the eight *naciones* already abide by this constraint, as Qulque Punku's left glacier tongue, which they used to ascend, has been melting so fast that their *ukukus* no longer can reach it. But even though the glacier's retreat is visible to everybody, and even though officials have tried to protect it for quite some time, many pilgrims continue to walk on it and cut off chunks of ice. Observing the habit of tying ice to dolls when she did fieldwork in the 2000s, Ceruti noticed an *ukuku* who "carried a doll, representing a miniature bear-man, on whose back he had placed a handful of ice, in the hope that he would be able to 'smuggle' the sacred cargo down the mountain, without catching the attention of the policemen" (2013, 5). She also observed pilgrims who had climbed the glacier. Ceruti offers the following report: "Bare footed, kneeling on the snow, they may pass hours watching the candles burn. Their silent prayer is combined with a careful contemplation of the flame, intended to provide them with answers to their most intimate concerns. Surely, they would have preferred to use larger candles, of the kind that the Lord of the Star of Snow is said to like best. But in recent times, they are only allowed to light small candles. Sometimes they cannot complete their traditional rituals[;] a policeman is likely to ask them to put the candles away, in a desperate attempt to stop the glaciers from retreating" (2013, 7).[13] Ironically, even though climate change threatens the pilgrimage, this enhances its effects. Using smaller candles may mitigate these, but they still do harm to the glacier just as cutting chunks of ice, setting off fireworks, or walking on it contribute to its melting. More bluntly, safeguarding the glacier and Quyllurit'i at the same time involves some very difficult compromises.

While not directly related to climate change, mining has confounded the organizers' perception of its impact on Quyllurit'i and emboldened followers' concerns about the pilgrimage site's environment. Mining has been going on in the area around Sinak'ara for some time, and from time to time the Ministry of Mining presents plans for new activities (Salas Carreño and Diez Hurtado 2018). After this happened in November 2015, Consejo de Naciones staged a protest in the city of Cusco on January 18, 2016, which I had the opportunity to join. Starting in the morning outside the ministry's office in Cusco, the protest culminated in the afternoon when more than a thousand people, of whom many were dressed as *ukukus* and in *comparsa*, defied the rain and gathered on the city's main square to listen to the Consejo's leaders, local authorities, and a representative from the Ministry of Mining. The anger triggered by the proposed plan of initiating new mining activities in the Sinak'ara area imbued the atmosphere and was ignited by the *comparsas'* music and the intense noise from the protesters' drumming, whistling, and yelling. The colorful and entertaining crowd showed that culture can be a powerful asset for making political statements.

In fact, the 2016 event was not the first time Consejo de Naciones had organized a public rally against mining. In 2007, as Peru's mining boom reached the Sinak'ara area, Consejo de Naciones also staged a protest in Cusco to denounce what it feared could turn into new more mining concessions. These were partly paused when the pilgrimage was declared intangible cultural heritage in 2011 and the Ministry of Culture established a protected area of 3,641 hectares for the shrine. But when small mining entrepreneurs ostensibly began to prepare activities along a ravine adjacent to the protected area in 2015, and rumors had it that illegal mining also was going on, Consejo de Naciones and the brotherhood thought it was time to respond by requesting the authorities to cancel these concessions and create a restricted buffer zone around Sinak'ara. And as their call for action had no effect, the two organizations decided to put pressure behind their demand by organizing the protest in 2016 (Salas Carreño 2020).[14]

Seemingly the initiative has had an impact; so far, no new mining operations have begun in the sanctuary area. However, according to the secretary of Consejo de Naciones the threat still exists. When I interviewed him after the protest in January 2016, he said: "The area around Sinak'ara is safe. They cannot do mining there. The biggest threat is against the area down to Ocongate where we walk over night after the celebration at the sanctuary." But the secretary exaggerates the danger, I was told by the woman (an anthropologist, by the way) in charge of managing Quylluri'ti's cultural heritage at the Dirección Desconcentrada de Cultura. "The protest is politically motivated," she explained to me in 2016. Even though the Ministry of Mining has issued the concessions to do mining within the area of the pilgrimage, the Ministry of Culture must authorize any new activities. She pointed out to me: "We'll never give such authorization. In fact, we're currently landmarking the 'sacred area' to protect it. And the nations know that." The woman also said that the pilgrims should be more worried about the neighboring communities that are predominantly Protestant and therefore oppose both Catholic and Andean religious traditions. Even though some of them, such as Mawayani, whence the pilgrims embark on the ascent to Sinak'ara, profit from the pilgrimage economically, they support the proposal to do mining in the area, hoping that it will create jobs and create new business opportunities.

In fact, according to the Dirección Desconcentrada de Cultura, the biggest threat to Quyllurit'i is not mining but climate change and the pilgrimage's own impact on the environment. In 2015 the woman in charge of its protection told me that the institute plans to restrict the growing commercial activities at Sinak'ara, particularly the many traders specializing in "fortune business," which it thinks is not essential to the pilgrimage. She also pointed to the "bank" selling forged banknotes and offering phony loans in the name of Quyllurit'i as examples of what the institute regards as activities that need to be reined in. "It is an important economic income for the brotherhood, but it is getting out of hand," the woman told me. Even more important is the effort to constrain the pilgrims' access to the glacier, mitigate their contact with the ice, and protect the creek's meltwater. "We must limit the number of people walking on the ice, prevent them from letting fireworks and putting candles on it, and improve the sanitary facilities," the woman said. On these issues Consejo de Naciones agrees. Its secretary pointed out to me: "We will do everything we can to protect Quyllurit'i's environment even if it requires changing our customs."

ADAPTATION COMPROMISES

The interviews I conducted with members of the eight *naciones* reveal that they are overwhelmingly concerned about Qulque Punku's receding glacier and that many wonder what will happen to the ice ritual in the future. Regretting but nonetheless consenting to the restrictions on the *ukukus'* performance on the glacier outlined by the Dirección Desconcentrada de Cultura, they fear the prospect of the future disappearance of all three glacier tongues, which will undermine the very idea of Quyllurit'i by rendering the ritual of initiating new *ukukus*, placing the cross on the glacier, and communicating with the ice inoperable (as already has happened to two of the eight *naciones*). Paradoxically, while the *ukukus* are restraining their ritual activities on Qulque Punku's glacier during the night, others are climbing it during the day. Thus, in 2015 I saw hundreds of pilgrims walking on Qulque Punku's central glacier tongue on the pilgrimage's principal day just as people told me they had seen pilgrims returning from the glacier with chunks of ice, which supports Ceruti's (2013) field observations in the mid-2000s. Reaching agreements with the eight *naciones* to adapt the cultural heritage of Quyllurit'i to climate change is one thing. Regulating the entrance to the glacier and controlling the individual activities of the growing number of pilgrims is quite another.

The secretary of Consejo de Naciones explained to me that in accordance with the Dirección Desconcentrada de Cultura the nations had agreed to limit the *ukukus'* access to the glacier. The agreement should be seen in the light of not only the retreating glacier but also the growing number of *ukukus*. The secretary told me that apart from the many new members the nations accept every year, Consejo de Naciones receives requests from people in Lima and Arequipa and even outside Peru to form new nations. "We appreciate that Quyllurit'i has become so popular, but we have to turn down these requests," he said. Rather than creating more nations, Consejo de Naciones needs to control the existing nations' activities. "We now have a total of 52,000 members of which 8,000 are *ukukus* or dancers," the man informed me. As already mentioned, three nations already refrain from physical contact with the glacier. "They don't have their own ice anymore," as the secretary phrased it. One of the five other nations has also imposed restrictions. Thus Quispicanchi, which is represented by the largest number of *ukukus*, only allows 40 of its 900 members to ascend the middle glacier tongue, while Canchis continues to let its 400 members do it. By the same token, Paucartambo's and Tawantisuyo's *ukukus* (350 and 600, respectively) are still allowed to use the right glacier tongue, which is the remotest and best conserved part of Qulque Punku's glacier. The biggest challenge, however, is not to restrict the *ukukus'* activities but to prevent other pilgrims from ascending the glacier during the daytime, ignoring the recommendations of the Dirección Desconcentrada de Cultura and Consejo de Naciones. "One year we formed a 'human wall' to prevent people from reaching the glacier. But they are just too many," the secretary told me. Salas Carreño, who has attended the pilgrimage several times in the past twenty-five years, affirms the secretary's consternation. He writes: "To charge the *ukukus* with the added task of closing all access to the glacier might take away too many *ukukus* and could be a source of conflict" (Salas Carreño 2021, 60).

The *ukukus* I interviewed, on the other hand, worried less about the pilgrimage's organizational and environmental problems and more about the constraints climate change and glacier retreat place on the annual rituals on the ice. One *ukuku* belonging to the Paucartambo nation said: "We go to the glacier to receive energy which lasts the entire year and gives us strength to be the guardians and transmitters of the faith and to connect the world above and the world of the earth." Then he explained how the ritual is changing form: "We ascend to the left side of the Qulque Punku glacier together with the Tawantisuyu nation at night and descend at 6:30 in the morning. We don't go up all the way as we used to; only a selected group from each nation do that carrying the cross." Another *ukuku* belonging to the Acomayo nation said: "We see the problem of global warming and that our glaciers and in particular Qulque Punku disappear with great concern. If they disappear our *apu* disappears too and we don't know what to do."

Some *ukukus* also worry about the remains of previous activities on the glacier disclosed by its retreat. They say they have found residues from former *ukukus* on the bedrock of the glacier as it is retreating. And what is more upsetting, the bodies of *ukukus* who have fallen into the glacier crevasses in previous years are now appearing. Many affirm that walking on the glacier at night is a dangerous business and that it is not unusual for *ukukus* to disappear, which some view as a sacrifice to the glacier and a symbol of the pilgrims' gratitude to its divine power (resonating with Ceruti's observations referred to earlier). An *ukuku* belonging to the Quispicanchi nation recounted: "Each year between four and five *ukukus* disappear in the ice. Some also die because the cold deepfreezes them." The man added: "Two years ago, the bodies of our disappeared brothers began to appear due to the thawing." To bury the bodies the eight *naciones* have made a graveyard next to Qulque Punku's receding glacier. According to the man quoted earlier, the graveyard serves as a site for the living *ukukus* to commemorate their dead predecessors. He told me: "Up to now we have buried around one hundred bodies." Some of the *ukukus* I interviewed wondered whether the emergence of the thawed bodies on the glacier bedrock erodes the very idea of reciprocity between the human and the sacred. One *ukuku* claimed that the appearance of dead bodies and the glacier's retreat are signs of a broken relationship. He said: "The ice eats us but then we eat it."[15]

But whereas many *ukukus* are troubled by the appearance of dead bodies on Qulque Punku's bedrock, they are pleased with the national and international attention that the event receives. An *ukuku* belonging to the Tawantisuyu nation said that UNESCO's declaration of Quyllurit'i as world intangible cultural heritage "is the biggest acknowledgment of our faith and culture." And when I asked him about the growing number of pilgrims attending the event, he answered: "I totally consent it. It shows the whole world knows about this acknowledgement." Several of the *ukukus* I interviewed affirmed this view, stating that Quyllurit'i's transformation into a global pilgrimage site makes them feel content. At the same time, however, many pilgrims worry about their own and other pilgrims' impact on Sinak'ara's environment and Qulque Punku's ice. One woman I talked with during the pilgrimage said: "There are not enough toilets here. We are just too many people." Another woman complained that some people harm the glacier. She told me: "I've seen a lot of people walking on the ice even though it's not allowed." She claimed that the growing number of pilgrims threatens the very idea of Quyllurit'i, which according to her is to pay respect to the sacred forces that protect the environment. Rhetorically, the woman asked me: "What's the point of coming here if we make the things worse, not better?"

Climate change is jeopardizing Quyllurit'i's cultural heritage but, ironically, the pilgrimage's growing popularity augments its effects. And even though UNESCO's declaration of Quyllurit'i as intangible heritage aims to safeguard the pilgrimage, it enhances the dilemma of protecting the pilgrimage's cultural tradition and the environment and in particular the glacier at one and the same time. Arguably, by drawing the world's attention to Quyllurit'i UNESCO has given the Ministry of Culture an almost impossible mission: to protect the pilgrimage's culture against its own success and to make it adapt to climate change without changing it. As my interviews with the people responsible for accomplishing this task show, the pilgrimage's stakeholders are aware of the harm it does to the environment and particularly the predicament it implies to conduct Quyllurit'i's ritual practices, as these contribute to the harm climate change is doing to a glacier. But while this is becoming clear to many of the pilgrimage's conventional followers, of whom quite a few are members of one of the eight nations and therefore can be considered "committed" Quyllurit'i pilgrims, a large number of "free-riding" pilgrims who attend on an on-and-off basis may not be informed and aware of the consequences their presence and activities have for the environment. While the authorities' and organizers' attempts to convey the message of climate change and environmental contamination to the "committed" followers have paid off to some extent, they still have a long way to go to reach the rest of the pilgrims and change their patterns of behavior.

The climatic and environmental challenge Quyllurit'i faces is similar to the one I described for Huaytapallana, though on a much larger scale due to different numbers of people attending the two events. Even so, the effects climate change and human activities have on Quyllurit'i's and Huaytapallana's ritual practices and religious expressions are very different. In both pilgrimages, environmental contamination, glacier retreat, and the appearance of former pilgrims' leftovers question the very idea of paying tribute to the mountain deities and the sacredness they represent. But while the offering constitutes a central ritual and is closely linked to the Andean image of the nonhuman agents' well-being in the Huaytapallana pilgrimage, Quyllurit'i is an inherently hybrid religious construct that draws on both native and imported notions of divinity. Created as a Christian adoration of the Savior in an Andean landscape of sacred mountains, Quyllurit'i's pilgrims celebrate both the legend of a Catholic image and the magic power of ice, snow, and water. Over time the syncretic event that emerged from these circumstances has altered ritual activities and shifted focus from one to the other religious world, which is reflected in the studies of Quyllurit'i in the past century. Thus, while earlier research foregrounds its legendary story about the Savior's revelation or the folkloric performance of its many dance and music groups, more recent studies (like mine) highlight the *ukukus'* relation with and interaction on the glacier.

This can be read as a "new age" trend, as Salas Carreño (2014) suggests, but it can also be seen as an ongoing shift of the pendulum between Christian and Andean religious worlds that is immanent in syncretic events such as Quyllurit'i and is contingent on environmental, political, and social changes.[16] If the *ukukus* no longer can ascend the glacier, the focus will simply shift to one of the pilgrimage's many other cultural and religious dimensions. In other words, Ceruti's warning that climate change represents a threat to Quyllurit'i or for that sake to Andean culture is not likely to materialize. In fact, phenomena such as climate change are precisely what make events such as Quyllurit'i and Huaytapallana so attractive to many people. Andean culture has always been on the move. What is changing is our notion of what is Andean. As more and more people from Peru's cities or even abroad take an interest in spiritual events and join Andean pilgrimages, adopting indigenous ideas of sacredness and spirituality—some driven by new age thinking, others by genuine worries about the planet's future—Andean becomes the label of a new cosmopolitical vision of how the world could look—even without glaciers.

Conclusion

The topic of this book has been two sets of relationships. On the one hand is the water-power complex and how the two intersect in the social and religious worlds of Andean people, water giving rise to ideas of domination and control and power interfering in and remaking the hydrological cycle. On the other hand is the climate-culture nexus and how the two mutually impact each other, climate change undermining Andean people's ritual practices and worldview accelerating the impact of global warming in the Andes. Investigating the ways these two sets of relationships unfold in a setting of chronic water scarcity, the book's aim has been to contribute to the current debates on how people account for and adapt to climate change and, particularly, how they make sense of their own roles in the climatic and environmental changes that disrupt their lives. In the following I first discuss the two relationships separately. I then reflect on what their convergence tells us about the future role of offering rituals in Andean people's struggles to access water, reinvent their livelihoods, and build new allegiances while reinvigorating their culture and reconfiguring their cosmology in an anthropogenic world. I conclude by making a call to anthropologists and other social scholars to engage in the effort to rethink the planet's future.

THE WATER-POWER CONUNDRUM

Climate change has irreversible consequences for Andean society and culture, causing glacier melt, water shortage, flooding, weather irregularities, extreme temperatures, and so forth, which is apparent in all four case studies. But these also show that Andean people take action to adapt to its impact by reinventing their coping strategies, seeking new alliances, and revisiting their ritual practices and worldview. Moreover, the water crisis has inspired the state to invest in infrastructural and developmental projects in the Peruvian highlands and engage in a new contractual relationship with the Andean people based on institutional cooperation and shared management of the country's water resources. In response to this opening, Cabanaconde has rearranged its water supply chain, and it now receives water from a state-built channel instead of the nearby mountain. The shift of water supplier that occurred at a time when out-migration of young members of the community to the United States had gained momentum has spurred the community to acknowledge the Peruvian state as its legitimate water governor and to introduce a state-designed water management model. It has also led to a cosmopolitical shift. While the community has made offerings to the mountains for centuries—in preconquest times even in the form of human sacrifices—they now pay tribute to the channel apart from paying water tax to the state and water tariffs to their user organizations. Nonetheless, the case study reveals that the water users' concept of power and accountability remains the same even though their change of ritual practices implies a shift of political allegiances. Water must be accounted for, whether to the mountain or to the state. In a similar vein, though different in form the tribute, the tax, and the tariff are all regarded as valid payments to smooth water metabolism.

Tapay's development points in another direction. Rather than turning toward the state for help, it tries to solve its water shortage by breaking a deal with a mining company that offers community members jobs and promises to construct a channel in exchange for Tapay's consenting to mining activities within its territory. Ironically, the company's commitment to enhance the community's water discharge is being made at a moment when the bulk of its able-bodied members have left for the cities and, consequently, many fields have been abandoned and the demand for water has declined. In contrast to Cabanaconde, which takes water from a state channel and routinely pays the water tax and tariff, many of Tapay's water users evade these. But while the community continues to conduct offerings to the mountain and the many springs, few water users expect the rituals will produce more rain. In their view the water crisis is a sign that the reciprocal relationship underpinning the offerings is breaking up. What remains is the mountain's raw power to punish the water users for their disrespect and their newly acquired consumer habits, which are commonly believed to be a threat to the environment. Many therefore fear that omitting the rituals will cause the anger of the mountains.

How the construction of a water channel to the community and the change of water supplier will affect its offering practice remains an open question. Even so, it is safe to conclude from the two case studies that external relations with the state and mining companies play a critical role in how Andean communities construe water metabolism and the power that drives it.

But climate change and the glacier melt and water shortage it causes do not only disrupt Andean cosmology by altering the terms of exchange with the mountains and disabling these in reciprocating the offering gifts, which Tapay and Cabanaconde illustrate. As Huaytapallana and Quyllurit'i demonstrate, climate change also generates social frictions by exposing the leftovers of previous offering rituals and causing environmental pollution because of the trash the followers of Andean pilgrimages leave on the ground. The two case studies document how the disturbing encounter with human residues on the pilgrimage sites and the dead bodies of former pilgrims leads to a change of power relations both in conceptual terms as the pilgrims become aware of the effects their activities are having on the mountains, and in political terms as the regional and national authorities begin to regulate the pilgrimages to mitigate their environmental impact. The growing political intervention in Peru's pilgrimages sheds light on not only how Andean pilgrimages become a battleground for social conflicts triggered by Peru's water stress but also how water scarcity challenges the pilgrims' ideas of mountains, glaciers, and ice as powers bestowed with metaphysical capacities.

Huaytapallana and Quyllurit'i both show that climate change poses a predicament for the organizers of Andean pilgrimages. The troubling news of glacier melt and water shortage has become a magnet for people from Peru's cities who look for answers to the country's environmental challenges and search for new identities at Andean sanctuaries. The surge of new pilgrims is welcomed by local shamans and pilgrims, who see it as sign of the growing recognition of Andean culture and religion. At the same time, however, the growing number of visitors accelerates the environmental impact of the ritual activities, contributing to the mountains' decay. Rather than invigorating the lives of the mountains, the pilgrims endanger them.

THE CLIMATE-CULTURE PUZZLE

In the heated debates about climate change, this is often represented as a wild beast and culture as its vulnerable victim, as an encounter between a global predator and local prey. This book shows that the relationship is more complex. Of course, in places such as the Pacific and Indian Ocean global warming and rising seas do have catastrophic sequences for not only people's culture and lifestyle but entire societies and nation-states. Likewise, in mountain areas, arctic regions, and tropical countries, climate change causes glacier lake outburst floods (GLOFs; as discussed in chapter 4), tsunamis, hurricanes, and other natural disasters that jeopardize people's livelihoods, sometimes their very existence. Yet my data suggest that even though climate change is real in the Andes and has serious consequences for the livelihoods and life-forms of Andean people, the latter experience it as one of many changes transforming their lives rather than as an isolated phenomenon. And even when climate change does impact their lives and livelihoods in visible and conspicuous ways, they perceive and account for it differently. In fact, the term *climate*, which has been part of Andean people's glossary for many years, is now itself an issue of dispute. In the vernacular climate is synonymous with weather, but the global and national discourse on global warming is gaining ground in the Andes, and people increasingly draw on the term *climate change* to explain the environmental changes they experience in their daily lives. And as the term disseminates it acquires new meanings.

As is evident from Tapay, some people inscribe rising temperatures, irregular precipitation, and water shortage in their collective memories and family histories and explain them as phenomena similarly to what their grandparents and other older relatives have related to them. In their view, climate change is cyclic and inherently natural; it comes and goes, and there is very little they can do about it. In a somewhat similar vein, many interpret climate change as the work of nonhuman forces. Peru's Evangelical movements believe climate change is God's will, and the idea that it is a punishment for humans' sins and lack of faith is widespread among their *hermanos*. By the same token, as Huaytapallana shows, some Andean shamans read it as a sign that the mountains are dying and that the world as we know it is coming to an end. Another variant of this line of thinking is the claim that climate change is a *pachakuti*, a turnaround of the world and the start of a new beginning, which resonates with Andean cosmology and its conception of history divided into pre-Inca, Inca, colonial, and modern epochs and its millenarian illusion of the return of preconquest time.

While religious or metaphysical readings of climate change create feelings of impotence, contrition, and in some cases hopelessness, its interpretation as a human-generated phenomenon can be the source of conflict, and just as the claim that climate change is anthropogenic can mobilize people politically, it may reinforce relations of inequality and hegemony. As Cabanaconde demonstrates, climate change is often explained as a local rather than global phenomenon; equally, it is common that people make sense of its consequences by pointing to their own rather than others' activities, which generates friction and feelings of blame and guilt. Comparing climate perceptions in the Alps and the Andes, Christine Jurt et al. (2015) reached a similar conclusion. Highlighting the symbolic boundaries and "us-them" divides that people affected by the consequences of climate change create to identify victims and villains, they found that while people in the Alps attribute climate change to tourists and other foreigners, Andean people ascribe it to their own activities. Their research leads Jurt and colleagues to conclude that people use their moral understanding of climate change and the identification of blame, loss, and harm it entails as hallmarks to create community identity and distinguish themselves from others. Arguably, this insight resonates with my third and fourth case studies of tensions between shamans, pilgrims, environmental activists, and regional authorities. Similarly, in their study of public climate perceptions in the Cusco region Adrian Brügger, Robert Tobias, and Fredy Monge-Rodríguez (2021) learned that more vulnerable groups (often characterized by being rural and having low income and education levels) tended to perceive climate change as "more consequential, closer, and as a more natural (vs. anthropogenic) phenomenon than those from less vulnerable groups" (Brügger, Tobias, and Monge-Rodríguez 2021, 1). They also found that people with Quechua as their first language and with the lowest education levels "were less convinced that the climate is changing globally and that people in the lowest income group were more convinced that it is changing locally" (2021, 21), reflecting the conclusions from the interviews I did in my first and second case studies.

It is not only the meaning of climate change that is difficult to establish in the climate-culture nexus. The import of culture can also be slippery. When climate is presented as a threat to culture, this is imagined as a static construct that people cling to as an emblem of their identity. But while the reification and copyrighting of cultural practices are important means of political struggles in some places, which is evident in my third and fourth case studies, in most parts of the world culture is better understood as the distinct way people do their daily chores, as I demonstrate in the first and second case studies. From this perspective, what is experienced as fixed and stable is actual change. More specifically, culture is a tool people use to solve the many challenges they face in their lives and the filter through which they make sense of the world and the problems and changes they encounter. It is not only climate that impacts culture; the latter also affects the former, which is why we talk of anthropogenic climate change. The question is not whether climate change undermines culture or whether human activities are causing climate change but the extent to which, on the one hand, people's cultural practices contribute to climate change and, on the other, people command the resources to adjust their way of life to the environmental change it generates. In fact, the answer to the global climate crisis is not that people try to safeguard their culture but that they adapt it to the climate change of today and make it instrumental to the effort of mitigating their own contribution to the climate change of tomorrow, a challenge displayed in the third and fourth case studies.

MOUNTAIN RITUALS AS
SELF-REGULATING THERMOMETER

But even though the popularization of mountain rituals and dissemination of Andean cosmology exacerbate the regions' environmental problems, they open doors for a new understanding of human agency and anthropogenic change, which my focus on the metabolic enactment that drives offering practices brings to the fore. To explore how Andean people adapt their water practices and ritual customs to climate change, I have drawn on the works of old as well as new thinkers who problematize the society-environment complex. To Marx, the main cause of not only social but also environmental injustice was capitalism. However, his theory about metabolism and alienation also addressed a more general aspect of the socio-natural order underpinning it: humans' separation from nature. Marx's thinking has helped me understand the social nature of Andean ritual practices. Particularly, it has been productive for my exploration of how the ritual enactment of water metabolism and the symbolic reassemblage of the offering items reconfigure the hydrological cycle and enable Andean people to engage in a relationship of exchange with the powers that control it. Likewise, the notion of alienation has inspired my reading of the ritual's changing meaning in a time of water stress. By uncloaking the alienation humans suffer from nature, climate change prompts Andean people to not only seek new political alliances and rearrange their water supply chains but also to reconfigure the symbolic metabolism that structures their interaction with the environment and recalibrate the offerings that mediate their access to water. In doing so, the Andean people recognize mountains and other planetary agents as part of both the second nature they and other humans have created and their own inorganic body and the culture it is part of.

Rappaport's theory, on the other hand, was influenced by cultural materialism and lacked a global power perspective, but unlike many of his contemporaries, who were material determinists, he viewed culture, and particularly rituals, as an important mediating factor between humans and nature. By highlighting the metabolic process that the offerings trigger, my study has evoked Rappaport's study of the Kaiko ritual and his suggestion that it contributes to environmental sustainability. Similar to Kaiko, Andean offerings present the society-nature complex as a relationship of exchange between humans and nonhumans who demand the former's respect in return for their use of natural resources such as pigs and water. But even though Kaiko and Andean offerings both serve as regulators of humans' interaction with the environment, their social dynamics differ. In Rappaport's account the ritual constitutes a self-contained cultural institution that ensures humans' access to meat without jeopardizing the environment. By contrast, I have described the ritual as a social practice that over time changes form and meaning, which has repercussions for our understanding of its social nature. In his later writings, Rappaport considered the limitation of self-regulating rituals such as Kaiko's warning that if they fail to adjust to the changing world, they may become maladaptive and create an environmental crisis. My data support Rappaport's claim that the regulating mechanism of offerings can be subject to change. However, they also show that the ritual serves as more than an instrument to safeguard the environment. It can be a vehicle for human agency and self-reflection. As an enactment of water metabolism, the ritual opens not only a line of communication with the metaphysical powers that control water and other natural resources but also a window on humans' own impact on the environment. If Kaiko is a thermostat to restore the human-pig ratio (Dwyer 1985), offerings and other mountain rituals are a thermometer that Andean people use to read the environmental effects of their activities. Metaphorically speaking, the thermostat in Rappaport's study automatically regulates humans' interaction with nature, while the thermometer in my study allows humans to self-regulate their agency by providing them with an updated input on how their doings change the water flow and the physical surroundings.

The melting of the Andes has serious consequences for the way Andean society and culture are practiced and imagined today, but it does not imply their disappearance. Even though most of the region's glaciers and ice are bound to disappear, the changing climate generates new economic opportunities, prompts new forms of cooperation, and opens the way for new forms of cultural creativity. As I have demonstrated, climate change incites the state and private actors to engage more actively in the highlands, invest in infrastructural and developmental projects, and forge new social contracts with Andean communities. Also concerned with global warming and environmental contamination are people from the country's cities, who in their search for new meanings of modern urban life take a renewed interest in Andean culture and review their perception of Andean culture and the perception of nature as an animated landscape. Climate change is therefore a pandora's box of losses and gains that jeopardizes the lives, livelihoods, and life-forms of many people but that also encourages them to rethink their ways of making a living, reinvent their cultural practices, and reformulate their claims to social and political rights.

My study is a call to anthropologists and other social scholars to engage in an anthropogenic world. By documenting the experiences of mountain pilgrims and other firsthand witnesses to climate change, anthropologists may act as cosmopolitical emissaries who translate the planetary disclosure of people experiencing glacier retreats into a language that is accessible to others and that may speak for the universal rights of all modes of existence. In such a mission, cosmopolitics implies not only the struggle of indigenous people and other marginal populations to mobilize nonhuman beings and other species as social and political agents but also the efforts of anthropologists to translate, communicate, and advocate alternative perspectives and unfamiliar worldviews as conceived and practiced by humans anywhere. Using ethnographic fieldwork in faraway places to challenge established ideas of the world has always been one of anthropology's finest goals. In a time of climate change and glacier retreat, this involves documenting how people who experience severe environmental degradation rethink the socionatural order of the world they inhabit. And perhaps even more important, it requires that we as scholars revisit our own role in the planet's metabolism and our contributions to a more just Earth system.

NOTES

INTRODUCTION

1. Rutgerd Boelens claims that "the Andean concept of *Pachakuti* is cosmic reordering" and that it implies releasing of built-up tensions through telluric and hydrological forces" (2015, 85). Peter Gose, on the other hand, reports that *pachakuti* is "a cataclysmic overturning and inversion of the existing order" (2008, 51) and a term that often is employed to describe sudden and unpredictable change in Andean mythology (Gose 2019).

2. Mountain regions are particularly vulnerable to climate change and water scarcity (Barnett, Adam, and Lettenmaier 2005). They cover 25 percent of the global surface and offer home and living space for 26 percent of the world's population (Diaz, Grosjean, and Graumlich 2003; Orlove, Wiegandt, and Luckman 2008). Mountain regions are also the world's major water towers, supplying half of its population with fresh water for irrigation, industry, domestic use, and hydropower (Beniston 2003). In arid and semiarid areas, mountains provide as much as 90 to 100 percent of the freshwater resources (Carey et al. 2017; Gagné, Rasmussen, and Orlove 2014; Vuille et al. 2018). But mountains are also some of the regions that are most vulnerable to climate change (Cebon et al. 1998; Drenkhan et al. 2015; Orlove and Guillet 1985; Yao et al. 2012), particularly in the tropics, where glacier melt is an urgent problem jeopardizing the future of the local population (Bradley et al. 2006; Carey 2010; Coudrain, Francou, and Kundzewicz 2005).

3. Peru alone contains 71 percent of the world's tropical glaciers; together with rain and ground water, these constitute its principal freshwater supplies. According to Peru's Ministry of Environment (Ministerio de Ambiente 2015), the country has already lost 22 percent of its glaciers and 12 percent of its freshwater volume (see also INAIGEM 2017). As the glaciers recede, Peru faces a major water crisis (Bury et al. 2011; Mark et al. 2017). The vulnerability of the Peruvian Andes is especially worrying because meltwater from its glaciers supplies not only Andean people but also the country's mining and agricultural industry and major cities with fresh water (Li 2015; B. Lynch 2012; Oré et al. 2009; Rangecroft et al. 2013; Vergara et al. 2007).

4. One of the ways people become aware of climate change and the threat it poses to human life is its effect on the planet's water supply. Fresh water makes up less than 3 percent of all Earth's water, and even though its deposits are renewable, the future prospect of clean drinking water for the world's population looks gloomy (Kundzewiz et al. 2008). Glaciers and icecaps that hold 68.7 percent of all fresh water and constitute the main water supply for nearly two billion people (or one-quarter of the current world population) are retreating, while groundwater, another important water source, is shrinking in many places (USGS 2018a, 2018b). As a result, the planet is experiencing an environmental crisis that puts water stress, water conflicts, and water justice at the top of the political agenda in both the Global North and Global South (Bakker 2007; Farhana & Loftus 2015; Paerregaard and Andersen 2019; Rodríguez-Labajos and Martínez-Alier 2015).

5. According to USAID's Climatelinks, Peru's climate vulnerability is linked to the country's high inequality and poverty, which is concentrated among rural, indigenous populations. More than 80 percent of the country's farmers practice subsistence agriculture that in some areas is rain fed but in others is irrigated, which generates increasing competition over water resources for consumption, agriculture, and industry. Glacier melt at accelerating rates enhances Peru's vulnerability. Other contributing factors are the risks the rising sea and extreme storms represent to the coastal urban population. Climate change is also expected to amplify natural disasters such as floods, droughts, and landslides, often triggered by El Niño. Finally, the environmental problems caused by agricultural expansion, deforestation, illegal mining, and air and water pollution exacerbate the climate risks Peru is exposed to (Climatelinks 2017).

6. According to Funtowitz and Ravets (1993), a postnormal science is defined by four criteria: (1) facts are uncertain, (2) values are in dispute, (3) stakes are high, and (4) decisions are urgent.

7. Mario Blaser writes about the concept of cosmopolitics: "This concept, first proposed by Isabelle Stengers (1997), differs from Kantian cosmopolitanism, according to which a cosmopolitan is one who rejects parochial allegiances and embraces the common world (the cosmos) as the grounding to work out differences among humans. In this conception the cosmos is transcendent and requires no discussion. What is debated (and has to be resolved) are the different views that, given their allegiance to their cultures and traditions, humans have about that cosmos" (2016, 546).

8. Drawing on Isabelle Stengers, de la Cadena (2010, 361) defines *cosmopolitics* as a pluriversal political configuration that connects different worlds within its socionatural, all with the possibility of becoming legitimate adversaries not only within nation-states but also across the world formations.

9. *Biological metabolism* is defined as "the totality of the chemical reactions and physical changes that occur in living organisms" (Cammack et al. 2008).

10. Karl Marx continues to inspire modern anthropologists; see Patterson (2009).

11. Karl Marx's notion of "metabolic rift" refers to the irreparable harm capitalist agricultural production causes to nature's own metabolic process.

12. The hydrological cycle is a widely held doctrine among physical geographers and other natural scientists. It claims that the world's water supply and water's movement on, below, and above the surface of the earth follow a so-called hydrologic or hydrological cycle. Implicit in this model of how water circulates, its volume remains constant even though the water takes different forms (ice, fresh water, saline water, and atmospheric water) according to climatic variables.

13. See Paerregaard (1987a, 1989, 1992, 1993, 1994a, 1994b, 1997a, 1997b, 1998, 2000).

14. See Paerregaard (2013a, 2016, 2020a) and Paerregaard et al. (2016).

15. See Paerregaard (2010a, 2010b, 2013b, 2014b, 2015a, 2017, 2018a, 2018b, 2019a, 2019d) and Paerregaard et al. (2020).

16. See Paerregaard (2018b, 2019b, 2020b, 2021a).

17. The video can be seen on YouTube at https://youtu.be/EymrvJ2O6V4.

18. See Paerregaard (2020b).

1. WATER, POWER, AND OFFERINGS

1. Glaciers are increasingly viewed as a symbol of climate change impact. In the words of Julie Cruikshank: "Concerns about global climate change are giving glaciers new meaning for many people who may previously have considered them eternally frozen, safely distant, and largely inert. Most of the world's glaciers now seem to be melting rather than reproducing themselves, becoming a new kind of endangered species" (Cruikshank 2005, 6).

2. My use of the term *cosmology* follows a tradition within anthropology to study the cultural worldview and religious universe of indigenous people (Descolá 1996). This approach examines cosmologies as toolboxes indigenous people use to make sense of the surrounding world and account for their own position in it. Common to such imaginaries is a configuration of the universe as a sociocentric model dominated by supernatural forces that live outside the social domain but nevertheless are recognizable to and in contact with humans, who are one among many modes of beings in a plural natural world (Descolá 2013). The key to a good and sustainable life in this model is to honor and pay tribute to the nonhuman powers, who in return offer humans protection and grant them permission to use natural resources such as land, water, and wildlife. Reciprocity and mutual respect between humans and life-forms are therefore essential to many indigenous peoples.

3. According to Xavier Ricard Lanata: "The *apu* is the spirit of the mountain (*urqu*): not in terms of the mountain being inhabited by the spirit but rather in the meaning that the mountain is an *apu*, a tutelary spirit, a master" (2007, 54).

4. In some regions of Peru, the *misti-indio* dualism has evolved differently. Rather than using the *indio* image to ostracize the rural peasantry, urban mestizos in Cuzco identified themselves as indigenous Peruvians and the proud heirs of the Inca culture and language (de la Cadena 2000). Ironically, while this has led to the revitalization of Peru's preconquest heritage and the introduction of indigenous

traditions and customs in public space, cultural prejudices and racial discrimination against the rural population are still widespread in the Cuzco region (Huayhua 2014).

5. See Paerregaard (2008a, 2012a, 2014a, 2015a, 2015b).

6. After many years of emigration and the reporting of Peruvians' experience as immigrants in other countries, Peru has become a country of immigration and finds itself engaged in a heated debate about the rights and wrongs of receiving refugees and migrants from other South American countries (Peru Reports 2018).

7. The creation of new jobs and the rising living standard that followed from this boom cut Peru's poverty rate by more than half (from 58.7% in 2004 to 20.6% in 2016) (World Bank 2018).

8. The law conceived of a new national system of water resource management, governing on three levels: (1) the national water authority, governed centrally from Lima; (2) fourteen administrative water authorities (Autoridades Administrativas del Agua, AAAs) to direct and implement water politics and legal norms regionally; and (3) several local water administrations that are responsible for overseeing the rights to and use of water. This system of governance is based on stakeholder participation and collaboration that encourage the water users to organize in water user committees at the community level (Comisiones de Usuarios) and water user organizations at the regional level (Juntas de Usuarios).

9. The Peruvian state has created water basin councils in several regions that gather all the water basin's stakeholders to discuss how water shall be managed (Paerregaard, Stensrud, and Andersen 2016).

10. The Juntas de Usuarios are nongovernmental organizations that represent their members' interests in the government's water institutions. The Juntas' members are the irrigation committees of the local communities. In return for the tariff the committees pay to the Juntas, they receive technical and logistical support (Stensrud 2021). The country's rural communities have been familiar with several of these measures because water tariffs were already introduced by the 1969 water law, but it has only been since the passage of the 2009 water law that the state effectively has begun to tax water.

11. In the wake of the 2009 law, the state has incited the communities' sense of water accountability by launching a campaign to educate them in proper care and use of water (Andersen 2019), fostering what Karen Bakker (2012) calls a biopolitical power of water.

12. Similar tensions in other regions where neoliberal policies aim to privatize natural resources have put the customary and often informal entitlements of marginal population groups to water under pressure and made the right to access, allocate, and consume water highly charged questions, politically as well as morally (Bakker 2012; Derman and Ferguson 2003; Farhana and Loftus 2015; Gálaz 2004; Groenfeldt 2013; Strang 2016).

13. ENCC was conceived of one year after Peru hosted COP15, the 15th United Nations Climate Change Conference.

14. The National Strategy for Climate Change.

15. Valderrama and Escalante report that during the offering rituals to the mountains, people use the term *mallku*, which otherwise is used to refer to Andean people's pre-Hispanic forefathers, to designate water (1988, 94–96).

16. The importance of water values and ethics is particularly evident in indigenous people's struggle to preserve collective rights to water sources and irrigation canals they have inherited from their ancestors and managed autonomously for centuries (Blaser 2013; Cruikshank 2005; de la Cadena 2015; Stensrud 2021).

17. Geremia Cometti reports that elderly members of the Q'ero community in Peru's southern highlands attribute the cause of climate change to younger people's neglect of conducting offerings and the rupture of the community's relation of exchange with the mountains (Cometti 2020a, 2020b).

18. Astrid Stensrud reports from her study in the Colca Valley that the aim of including seawater in the offering is to *llamar al agua*, "a technique that performs the hydrological cycle in order to call the water from the ocean and make clouds and rain" (2016a, 13)

19. The idea of smoothing the hydrological circle taps into a cosmology that presents the earth as floating on a great sea that connects all parts of the world and that attributes the control of the water flow to main characters in the Andean pantheon: *viraqocha* (the god of creation), *mamaqocha* (the goddess of the sea), and *pachamama* (the goddess of the earth) (Sherbondy 1982; Urton 1981). The cosmology draws on the Andean notion of cyclical time, which raises the question of how the four versions of water metabolism proposed in this book rely on and/or produce different temporal assumptions and rhythms. I discuss in more detail in what way the intersection between climate change, mountain offerings, and changing ritual performances is linked to the temporality in the current climate crisis elsewhere (see Paerregaard 2023).

20. Other terms for offering are *alcanzo* (Spanish: presentation) or *despacho* (Spanish: farewell) (Paerregaard 1989).

21. Districts (*distritos*), which are subdivided into annexes (*anexos*) and population centers (*centros poblados*), are the smallest unit in Peru's administrative hierarchy. The districts of Tapay and Cabanaconde are both part of the *provincia* (province) of Caylloma, which again forms part of the *departamento* (department) of Arequipa.

22. Formally, the state began to charge tax on the villagers' water use in 1994 (though with little success), and with the 2009 water law it has also authorized the Juntas de Usuarios to collect a water tariff for the maintenance of their members' water infrastructure. However, as I discuss in chapter 2, not all water users pay the tax and tariff.

2. TAPAY: THE OFFERING MUST GO ON

1. In the first century after the Spanish conquest, Tapay represented only 4 percent of the region's tax-paying population, but during the colonial period the district

grew faster than other Colca settlements, and in 1791 it made up 8.1 percent of their total population with 1,215 individuals, including tax-paying as well as tax-exempt residents (Paerregaard 1997a, 36–40). Tapay continued to grow in the early republic period. According to a census conducted in the Colca region in 1843, Tapay's population was 1,459, of which 638 were males and 821 were females, representing 7 percent of the total Colca population. And according to Peru's first national census in 1876, it had 1,509 residents, which made it the fourth largest settlement of the Caylloma province, constituting 7.9 percent of its population. The next national census, in 1940, showed that Tapay grew further in the first half of the twentieth century, when it reached a demographic peak of 1,580 residents.

2. As a district in Peru's administrative hierarchy, Tapay elects its own mayor (*alcalde*) as well as representatives of Peru's judiciary power (*juez de paz*), while the representative of Peru's executive power (*gobernador*) is appointed by the province's *sub-prefecto*, who again is appointed by the regional government's *prefecto*. The three authorities, which also are represented at the subdistrict level in Tapay's three *anexos*, act as servants of what Colloredo-Mansfeld (2009) has called *vernacular statecraft*: a body of bureaucratic offices and administrative duties that are institutionalized by the state but managed by indigenous laypersons and embedded in local practices.

3. Tapay experienced a population decline, from 1,404 individuals in 1961 to 1,239 in 1971, 930 in 1981, and 820 in 1993 (Paerregaard 1997a, 39). In 1986 I conducted a census in Tapay that showed the district had 983 inhabitants, which is more than Peru's national censuses from 1981 and 1997 counted (1997a, 39). The difference can be explained by the methods applied. While the census counters only knocked on doors and asked for the number of people living in the house, I registered every single inhabitant by name, age, gender, family relation, place of birth, and other characteristics, a method the national censuses of 2007 and 2017 also applied.

4. Compared to other Colca villages, the scale of Tapay's outmigration and the demographic decline it has entailed is remarkable. Between 1876 and 2017 the overall population of the Caylloma province grew from 19,111 to 67,745, whereas Tapay's population dropped from 1,509 to 449, which means that it is now the smallest of the province's nineteen settlements and constitutes less than 1 percent of its population.

5. My 1986 census showed that women constituted 52 percent and men 48 percent of the district's 983 residents. Thirty-one years later, the 2017 census showed that 228 of the district's permanent residents were men and only 221 women.

6. While four out of ten villagers were between one and eighteen years old in 1986, in 2017 this age group only comprised 25 percent of the population. Moreover, while 49 percent of the population were between nineteen and sixty-four years old in 1986, villagers age sixty-five or older constituted 26 percent of the population in 2017. Unfortunately, the 2017 census only provides information on Tapay's population's age distribution that comprises the mineworkers. But in 2011 I conducted a survey among twenty-four households in Tapay (approximately 12% of all households) that included questions on the permanent residents' gender and age. It showed that while 38 percent of the household members were between ages eighteen

and forty-nine, 32 percent were younger than eighteen and 30 percent were age fifty or older.

7. My 2011 survey showed that the villagers' median age was 35.5 years, and that the average number of Tapay's households is 3.04.

8. The hamlets are Tapay (the district capital), Puquio, San Juan de Chuccho, Cosñirhuar, Malata, Paclla, Llatica, Fure, Tocallo, and Puna Chica.

9. In the community of Lari water is transported from sources above the village via three separate canal systems that feed into the irrigation system (Guillet 1992). Thawing snow from the surrounding mountains also contributes water to Yanque. However, irrigation in the community of Yanque differs from that of Lari, in that it is supplied by two different mountains sources, one on each side of the Colca River (Brandshaug 2019; Valderrrama and Escalante 1988). Consequently, each moiety of the village controls its own irrigation system. Irrigation in Cabanaconde, which I discuss in chapter 4, largely adheres to the pattern of Lari, disposing of three main water sources, but with only one local spring, used mainly for drinking water (Gelles 2000).

10. The idea of the 1969 law was to create a common institutional framework for all Colca districts and to link their irrigation committees by transforming them into subdivisions of a regional Junta Directiva de Regantes (a forerunner of today's Junta de Usuarios; see chapter 1).

11. With the 2009 law the name *comisiones de riego* was changed to *comités de usuarios de agua*.

12. *Brinco* is also used in other Colca districts, including Yanque, though under different names (Brandshaug 2019; Paerregaard, Ullberg, and Brandshaug 2020). The idea is that the *regidor* makes a list of water users within the irrigation cluster who have requested water. The governing principle is that the water user who sows his/her crop first is entitled to receive water first.

13. As in most Colca districts, Tapay's water supply only allows for four irrigation rounds in the dry season, called *qelhua, hualtay, hallmay,* and *melka*.

14. Even though Tapay's principal irrigation clusters are now organized in user committees according to the 2009 law, informal water groups continue to manage Tapay's smaller water springs. Consequently, these are not formally represented at Junta de Usuarios meetings and do not have right to its services. Most users of Tapay's springs, however, own fields in other parts of the district and are therefore members of one of its four user committees.

15. To determine the water tax and water tariff, the Ministry of Agriculture has conducted a PROFODUA (Programa de Formalización de Derechos de Uso de Agua) in most Colca districts, a program that issues a license to use water to irrigate and register the local water users. In several districts the program has been met with resistance (Vera Delgado and Vincent 2013, 198). So far, however, no PROFODUA has been conducted in Tapay.

16. Tapay's water shortage is difficult to estimate. Many villagers report that the district's rivers and springs yield less water than they did twenty to thirty years ago. They point to Tapay's half-empty reservoirs as evidence. These used to be filled in twenty-four hours; now it takes twice as long. However, shortage of water is a relative

phenomenon that depends on the number of people who demand it and on the eyes that see it. Due to outmigration, about one-third of Tapay's fields have been abandoned. Some claim this is because there is not enough water. Others say it is proof of Tapay's massive outmigration and a sign that the demand for water has diminished.

17. *Pago* refers to a ceremonial arrangement involving a packet of offering items, often referred to as a *mesa*. *T'inka*, on the other hand, is a common denotation for all kinds of offerings, including daily informal libations to *pachamama* (Mother Earth), *gentiles* (remote forefathers), and *machus* (mountains), as well as to animals, houses, and so forth.

18. Andean anthropologists have discussed how the merging of the celebration of the dead on November 1 and the beginning of the rainy season associates water with the afterlife; see Harris (1982) and Paerregaard (1987b).

19. Even though not everybody agrees, many villagers believe that Seprigina is female, which is unusual in the Andes. By contrast, Hualca Hualca, the mountain of neighboring Cabanaconde, is often referred to as male.

20. Villagers use different names to refer to the bugs: *cocoyero*, found in all crops; *mosca minadora*, found in broad beans; *chocolatada*, which damage potatoes; and *shilwe*, which attack corn.

21. These programs include Qaliwarma, which offers food to children in low-income households; Juntos, which helps single mothers with school-aged children economically; Pensión 65, which supports people over age sixty-five who don't have an income; FISI, which provides cooking gas to poor people; and SIS, which is health insurance for the poor.

22. The owner of the mine called Tambomayo is Buenaventura, Peru's biggest mining company.

23. The mining company's promise is to supply Tapay with two hundred liters per second, which the villagers plan to divide into four parts of fifty liters per second each: one for the main hamlet, Puquio and Chuccho; another for Cosñirhua; a third for Malata; and a fourth for the newly formed settlement of Belén, which is populated by villagers from the hamlets of Llatica, Furi, Tocallo, and Paccla. Compared with the district's current total water discharge of less than one hundred liters, the supply will make a significant difference.

24. The villagers have produced *palta nativa*, a local avocado, for a long time and prefer it because it can yield fruit year-round. The national and particularly the export market, however, demand Haas (or Fuerte), which is smaller and more watery, and which can be stored for a longer time.

3. CABANACONDE: THE HOLE IN THE CHANNEL

1. Auto Colca is a semiautonomous organization that promotes tourism in the Colca Valley. Its main function is to charge tourists traveling in the region an entrance fee and use it to improve roads, establish tourist facilities, built museums and other cultural sites, and so forth.

2. Envisioning a development project to support small-scale farming in the early twentieth century, the military government of Juan Velasco Alvarado (1968–1975) created what at the time was one of the world's most expensive irrigation projects through public (35%) and international (65%) funding (Stensrud 2016b, 2021). From its origin, the Majes project was conceived to have two stages. The first stage (Majes I) was built between 1974 and 1982. It consisted of the Condoroma Dam, which is located in the highlands and fed by the nearby Colca River, several intakes, and 101 kilometers of tunnels and canals that lead the water through the Colca Valley, where it supplies the communities on the west bank, to the Majes and Siguas plains on the coast, where it irrigates 15,000 hectares of fields that produce a variety of crops such as alfalfa, which is used as fodder for beef cattle and dairy cows in the region; potatoes, corn, and legumes for the regional market; and *aji* pepper and artichoke for export (Paerregaard, Ullberg, and Brandshaug 2020). The second stage (Majes II) is currently under construction (Ullberg 2019).

3. According to Gelles, of a total population of 634 in 1790, almost a quarter (145) were Spaniards or mestizos (2000, 32).

4. In Peru the peasant community is a legal social entity that collectively manages natural resources such as land and water. Membership implies not only rights to land and water but also obligations to do communal work and often has strings attached, such as local descendance and residence.

5. In 1876 when Peru conducted its first census, Cabanaconde was the most populated district in the Caylloma province, with 2,499 residents, almost double the size of Chivay, the province's current capital and now the biggest settlement. In the twentieth century Cabanaconde grew from 2,960 in 1940 to 3,363 in 1961, then to 3,368 in 1971, and reached 3,373 in 1981, after which the population started to decline (Paerregaard 1997a, 39). In 1993 it fell to 3,196, a trend that has continued through the present.

6. In 2017 there were 925 men and 940 women in Cabanaconde. Twenty-eight percent of Cabanaconde's population is between one and eighteen years old, while 17 percent is age sixty-five or older, and only 55 percent are between nineteen and sixty-four years old. The section of Peru's total population that is age sixty-five or older only constitutes 8 percent.

7. Migrants also make individual donations to Cabanaconde. One migrant told me that every year the students finishing *colegio* (Spanish: secondary school) in Cabanaconde ask him to sponsor the trip they make to celebrate their graduation. Traditionally, students traveled to Lima or Cusco for a couple of days, but he said now they ask for many thousands of dollars to travel to the United States: "For several years I agreed and sent them the money but when they asked for money to go to the US I stopped. They don't want to travel to celebrate their graduation. They want to migrate."

8. There are signs that Cabanaconde's US-bound migration is slowing down. A woman who graduated from Cabanaconde's high school in 1998 told me that sixteen of her thirty-two classmates had migrated to the United States. In 2019, on the other hand, a villager said that today only a few of Cabanaconde's high school

students migrate to the United States, not only because traveling is much more hazardous but also because there are more opportunities in Peru's cities.

9. The hotel, which is located on the highest point in the settlement and therefore can be seen from far away, has been under construction for almost ten years. Some villagers claim it is located on a *huaka* with bones from *gentiles*, which brings bad luck.

10. Migrants also celebrate the Virgin del Carmen in Washington, D.C., in a cave called the Gruta de Lourdes on the city's outskirts. The event, which takes place in July, is today the single most important migrant gathering in Washington. As in Cabanaconde, a *devoto* assumes the responsibility of organizing and financing the event, which includes a religious ceremony followed by a party with food, music, and dancing.

11. According to another villager, the *devoto*'s brother proclaimed during the event that he intended to sponsor the fiesta in 2015. Yet when the man realized that many locals disparaged his plan, he withdrew the offer, saying: "!Se cargaron!" (They fucked up!) As no one volunteered to sponsor the fiesta the following year, Cabanaconde made a so-called *junta* (public fundraising) to organize the event, which many found more enjoyable than the previous fiestas. However, in 2017 a migrant from Lima, and in 2018 and 2019 migrants from the United States, sponsored the fiesta, adding more fuel to the competitive spirit that drives it. Another indication that fiestas will continue to be a showcase for migrants to demonstrate their wealth is that they are now the main sponsors of the celebration of Candelaria, Cabanaconde's second most important saint, in February also.

12. Economically, the villagers belong to the poorest segment of the Peruvian population. Their average annual income is US$976, and one-third do not have a regular income at all. Only 30 percent make more than US$1,667 a year, which indicates that to a large extent the households are self-subsistent.

13. The survey comprised one hundred households, representing 18 percent of all households in Cabanaconde. The research questions were as follows: Who are the household members (age, gender, kinship relations, religion, place of birth and residence)? What are the household's landholdings (dry land, irrigated land, pastures etc.)? How many and what kind of animals does it have? What is its annual income? What are its primary and secondary livelihoods? What crops does it grow? What is its use of water (irrigation, drinking water, domestic use etc.)? What is its migration history? What is its status as a community member?

14. Other sources of income included trade (5%), tourism (4%), construction (4%), technical consultancy (2%), and handicrafts (2%).

15. The survey also showed that the villagers' average age is 38.5 years, and that the average household has 3.1 members.

16. Since 1983, the Majes Project has been administered by Autodema (Autoridad Autónoma de Majes, Majes Autonomous Authority), which was an autonomous public agency under the Ministry of Agriculture and the central government until 2003, when it passed to the regional government of Arequipa. Autodema's public experts will continue to manage the project until the construction of its amplification

(Majes II) is completed and the private consortium that is building it takes over and runs the infrastructure for the sixteen years of its concession.

17. The water committees are organized in four water user groups, including the Junta de Usuarios del Valle de Colca, to which both Cabanaconde and Tapay belong. Based in Chivay, the Junta is a private nonprofit association that represents a total of nine thousand water users organized in forty-six water committees (divided into fifteen *comités de usuarios* and thirty-one *comisiones de usuarios*) in the Colca region. More than half of these committees take water from the Majes channel.

18. For more details about the case, see Paerregaard, Ullberg, and Brandshaug 2020.

19. My attendance was facilitated by Susann Ullberg, who at that time worked with me on a research project on the Majes Project.

20. According to Vera Delgado and Vincent, the irrigation supply registered for the Colca Valley districts "may be related to the design discharge of local canals or offtakes (the discharge capable to supply the area to be irrigated in a given time period, including allowances for water losses)" (2013, 204).

21. My participation in the trip was facilitated by Malene Brandshaug, who at that time was my graduate student doing fieldwork in Yanque. My description of the trip draws on Brandshaug's dissertation work.

22. Following are the yields of the five valves: Villa Colca, 80 liters; Media Luna, 70 liters; La Campiña, 205 liters; Joyas, 55 liters; and Castro Pampa, 120 liters. Hualca Hualca also contributes to Cabanaconde's water discharge with 200 liters, of which 55 liters go to Joyas and the rest to La Campiña. The latter, however, only receives 40 liters, as the remaining 105 liters are lost. Moreover, Cabanaconde has four springs. Puquio Pampa supplies the district's drinking water, while the other three springs (Taqrari, Granadillasnero, and Jolpa) are unused. Together they yield 40 to 50 liters.

23. Community members cannot sell the land but have the right to pass it on to their children. In 2016 the total number of community members was 549, of whom approximately 50 do not have communal land. Another 20 to 30 are waiting to become members, which requires five years of residence in Cabanaconde. A total of 30 members were born elsewhere. Some villagers say migrants should be denied the right to use communal land. They argue that rather than allowing migrants to rent the land to outsiders, the community should pass it on to the villagers on the waiting list.

24. One of the reservoirs, Joyas, was built by a government program called PRONAMACH in 2003. It only stores water from the Hualca Hualca River. Another, Puquio Pampa, Cabanaconde's oldest reservoir, is fed by water from both the Hualca Hualca River and a local spring. A third is Culcurumi, which was built by the government in 2008. It stores water from the canal that runs through the settlement and that is fed by both the channel and the Hualca Hualca River. A fourth reservoir is Media Luna, which takes water from the channel and was built in 2000 by a program run by Peru's Ministry of Agriculture called PSI. The same program also built Auqui Pata, which takes water from the channel. Finally, in 2003 the

regional government built Villa Colca (also known as 18), which takes water from the channel.

25. Even though the project is still in the initial planning phase, Majes II has already aroused the anger of the population of the neighboring region of Cusco, who say it will tap water from their water sources. In Arequipa, on the other hand, the affected settlements see the project as a future provider of water. Among these settlements is Cabanaconde, where several community leaders I interviewed told me that they plan to make new claims to water from the Majes channel when the project has been completed, just as the village did in 1983. One leader pointed out to me: "Majes goes through our community and when they enhance the water flow, we have the right to more water." He even named the areas that Cabanaconde plans to irrigate with the water.

26. In an epilogue Gelles (2000, 163), writes that by 1997 Cabanaconde had in fact adopted the state model.

27. Arguably, such claims are problematic because, on the one hand, they erase Inca imperial violence, attributing it to the local community, and on the other, this type of Inca-imposed sacrifice was about not only water provision but also the reproduction of imperial power.

28. In 1986 an engineer who had worked in the Colca Valley for many years told me that he had known an old man in Pinchollo, Cabanaconde's neighboring settlement, who recalled how the community once sacrificed a young girl to Hualca Hualca, its principal water source at that time. Interestingly, the man didn't describe it as a sacrifice. In his account, the girl jumped into the mountain of her own free will.

29. In his review of the precolonial history of the Colca Valley, Noble David Cook, writes that the two ethnic groups inhabiting the region practiced different forms of skull deformation. While the Collaguas formed the skull as a cone to match the shape of a volcano, "The Cabanas [the people of Cabanaconde], in contrast, strove to flatten and widen their heads. The great mountain of Hualca Hualca is not a cone at all, instead it takes the shape of rugged blocks. Not surprisingly, the Collaguas told the Spaniards that they believed that their Cabana neighbors, by widening their skulls, were 'very ugly and disproportionate'" (Cook 2007, 13).

30. Even though Cabanaconde continues to make offerings to a number of intersections and connecting points in its irrigation infrastructure, several of the presidents of Cabanaconde's irrigation committees explained to me that while participation in the tribute ceremony to Hualca Hualca previously was viewed as a way the villagers affirmed not only their community membership but also their water rights, these are no longer linked to ritual offerings whether conducted collectively or individually. Instead, they pointed out, water rights are now obtained by buying the ticket from the irrigation committee and paying the tax and tariff to the Junta de Usuarios in Chivay.

31. My survey resonates in broad terms with the study of climate perceptions in a village in the Cusco region by Brügger, Tobias, and Monge-Rodríguez, who also found that water was an issue of concern: "Water-related problems were at the

forefront of respondents' minds. Not only did more than one-third (37.2%) of the sample spontaneously mention water scarcity as one of the most important future threats to their country, but many respondents had already personally experienced negative consequences arising from either water scarcity, droughts, or flooding. Respondents also directly linked water-related problems to climate change and expected them to intensify in the future" (2021).

32. The villagers of Tapay and Cabanaconde cannot account for the origin of the different irrigation methods the two communities apply. I attribute them to the communities' geographical settings and water supplies: Tapay's mixture of fields situated on extremely sloping terrain and fed by water from several rivers and springs versus Cabanaconde's La Campiña, which is relatively flat and fed by only one water source (Hualca Hualca).

33. An alternative method is drip irrigation, which involves dripping water onto the soil from a system of plastic pipes fitted with outlets. An even better method is surface drip irrigation, which reduces soil water evaporation losses by applying water directly inside the ground instead of on the surface. The problem with both methods is that they require investment capital and knowledge of how to apply them—both scarce in Cabanaconde.

34. The change from traditional crops for self-subsistence to cash crops and livestock breeding, combined with migration and alternative incomes in response to climate change and globalization, is seen in other parts of the Andes too. From their study of the Peruvian community of Langui, Lennox and Gowdy conclude: "The transition towards livestock-based agriculture, and towards migration and non-agricultural employment occurring in Langui, Peru shows a great capacity of individual households within the region to adapt to the challenge of climate change and economic globalization that they face" (2014, 161).

4. HUAYTAPALLANA: THE *APU* THAT IS DYING

1. The aim of this event is to celebrate Santiago, a Catholic saint believed to protect the animals. Today the event gathers thousands of people, the majority of whom come from Huancayo and other urban centers.

2. Male villagers from Acopalca have worked as sheepherders in the Western United States for several decades, a migration practice that is common in the region's high-altitude communities; see Paerregaard (2010b, 2011, 2012b).

3. Local scholars have found that while currently 93.7 percent of Huancayo is situated in zones of medium water vulnerability and only 2.3 percent is in low water vulnerability, by 2030, 33.7 percent of the city will be in high-risk zones (Gómez and Santos 2012, 25).

4. SEDAM's complete name is Empresa de Servicios de Agua Potable y Alcantarillado Municipal (Municipal Agency to Service Drinking Water and Sewerage).

5. The sources of Huancayo's water supply vary during the year. In the rainy season from January to April the Shullcas River and underground water each provide

50 percent of the supply, while during the dry season 60 percent of Huancayo's water supply comes from glacier and rain lakes (the glacier lake of Lazuntay alone provides 20 percent of Huancayo's total supply), 30 percent comes from underground water, and the rest comes from eighteen springs in the surrounding area.

6. When I interviewed two engineers from SEDAM in 2017, they told me they were working on two alternative projects to solve Huancayo's water crisis. One is to transport water in tubes from Pachacayo (80 km northwest of Huancayo), the other to take water from a spring in Sicaya (15 km northwest of Huancayo). Both projects, however, are not only costly but may add fuel to the competition over the region's depleting water resources.

7. JASS stands for Juntas Administradoras de Servicio y Saneamiento (associations of service and sanitation managers), which are rural community-managed water and sewage service organizations.

8. The movement is led by Huancayo's bishop, Pedro Barreto, who in an interview I did with him in 2015 said he had received death threats for his role in the fight against the contamination of the Mantaro River; see Graeter (2017).

9. In the Andes chewing coca leaves is a social habit that people practice daily, but at offering ceremonies they carefully select the intact rather than the broken leaves, as these are viewed as essential gifts for the mountains and other nonhuman powers.

10. People told me that famous singers such as Sonia Morales and Dina Paúcar have visited Huaytapallana to ask for luck in their careers.

11. According to the regional government, the trash comprises broken bottles (60%), the remains of burned candles (10%), plastic (10%), and miscellaneous (20%).

12. A legal construction that was introduced by the Fujimori government in the 1990s for areas of regional protection was reinforced in 2005 with the aim of establishing protected areas on state-owned land that is of local or regional importance. Once established, the areas are often managed by the regional governments, as happened in the case of Huaytapallana (Haller and Córdova-Aguilar 2018, 60).

13. The director's Spanish title is Gerente Regional de Recursos Naturales y Gestión del Medio Ambiente de Junín.

14. Pedro's view may sound somewhat extreme (and is certainly provocative), but it does point to an important predicament in Western globalist ideas of climate change. Thus, recent research highlights the contextuality of people's perceptions of climate change, which are shaped by their own experiences and cultural ideas, and therefore is at variance with the science-based and Western-generated global discourse on climate change (Crate 2011; Mathur 2015). As Jasanoff (2010) points out, climate science tends to separate the epistemic from the normative and detaches global facts from local value, destabilizing knowledge at the same time that it seeks to stabilize. She continues: "It thus cuts against the grain of ordinary experience, the basis for our social arrangements and ethical instincts" (2010, 237).

15. Pachacámac is a small town on the Peruvian coast just south of Lima. During the Inca empire it was an important sacred site.

16. Pedro was one of the three men who spoke at the ceremony in 2014. The other ritual specialists were the leader of a bilingual Spanish-Quechua school in

Huancayo and a renowned intellectual who owns an accountant office on the city's main street.

17. Arguably, while Pedro transgresses the notion of purity, other *layas* and many of his followers objectify it by promoting an image of Huaytapallana's ice as "pure nature" that epitomizes the sacred (Gremaud 2014). My travel companions' (and my own) dissatisfaction with Huaytapallana's contaminated air and environment, then, was driven by the illusion of pure nature and the denial of the dangers that threaten the symbolic borderlines between purity and pollution but that allow us to live with ambiguity and the compromises required in life. Yet even though many visitors crave safety in the form of pure nature, their encounter with Huaytapallana and its threatened environment erodes the illusion of purity and provides them with a new perspective on the relations between humans and their surroundings.

18. The director in charge of protecting Huaytapallana's environment explained to me that people are allowed to bring beer, wine, and alcohol to the mountain. It's the bottles that are prohibited, she pointed out. She said people are encouraged to bring liquids in *materiales artesenales* (Spanish: artisanal materials), containers made of organic material. Music bands that traditionally are an essential component of offering ceremonies in the Andes are also banned because the sound may provoke the breaking up of the glacier.

5. QUYLLURIT'I: THE GLACIER THAT SHINES LIKE A STAR

1. I spell the pilgrimage "Quyllurit'i" following the recommendation of late Peruvian anthropologist Jorge Flores Ochoa, who suggested that the name's spelling should follow vernacular Quechua pronunciation (Salas Carreño, personal communication).

2. The rebellion was an uprising against the abuse of not only the Spaniards but also the *kurakas* (Quechua: native leaders), and it divided the native population as much as it united them (Sallnow 1987, 213). See also Szeminski (1984).

3. Some of these processions now have a worldwide extension; see Paerregaard (2008b). For an overview of Peru's many pilgrimages, see Olivas Weston (1999).

4. Even among the neighboring indigenous communities, different interpretations of the legend prevail, which Sallnow observed. He reports: "Most conspicuously, however, in contrast to both the official account and that from Ccata, the Qamawaran version males no mention whatsoever of priests or mistis. The cult is represented here as an entirely Indian affair. The original witnesses to or participants in the miracle are a pair of Indian boys; thereafter, devotees in traditional nación groups come to dance in eve increasing numbers before the taytacha" (1987, 211).

5. Elaborating on this dualism, Sallnow writes: "This dualistic design casts the entire congeries of pilgrims as two great moieties, labelled 'Paucartambo' and 'Quispicanchis', respectively. These are the names of the adjacent provinces whose mutual boundary passes through the Sinakara mountain range" (1987, 217). He

continues: "In fact, the labels are synechdochic. 'Paucartambo' in this context also includes the provinces of Cusco, Calca, Urubamba, and beyond—that is, the predominantly agricultural zone to the northwest of the shrine, the zone stretching toward the tropical forest; 'Quispicanchis' also includes the provinces of Acomayo, Canas, Canchis, and beyond -that is, the mainly pastoral zone to the southeast of the shrine, the zone stretching toward the tundra. This sociogeographical division is amplified still further to connote the ecological contrasts of valley versus mountain, lowland versus highland, and—polarized to the extreme—jungle versus puna. Ethnically, it comes to stand for the linguistic and cultural divide between Quechua and Qolla (Aymara), which in turn becomes transformed into the critical cleavage of Indian (native) versus mestizo (Spanish)" (1987, 217).

6. The *ukukus'* apparently marginal position in the pilgrimage should not be mistaken for a lack of dancing talent. As Poole writes: "The ukuku's clownish antics and prankish mimicry are used to emphasize his status as the most agile, skilled dancer of the comparsa" (1990, 99). Just to mention a few, *comparsas* comprise such figures as the *kapac qollas*, the *ch'unchos*, the *majeños*, and the *awqa chilenos*, who represent, respectively, wealthy traders from Qollasuyu (Puno), jungle Indians, nineteenth-century alcohol merchants from Majes (Arequipa), and Chilean soldiers from the nineteenth-century War of the Pacific (Poole 1990).

7. Poole reports that *comparsas* are widespread in the Cusco area, where "comic ukuku-like characters appear dressed as medics, lawyers, or even tourists." She adds: "The prominence accorded such "outsider" figures in the Cusco dance repertoire suggests that their role is, in fact, quite critical to the choreographies of transformation, devotion, and ritual identity spelled out in dance" (1990, 106).

8. Poole portrays the *ukuku* as the perfect hybrid: "He is half-animal and half-human, half-policeman and half-clown, half-boy and half-man. His role in fiestas is simultaneously as a keeper-of-order and a disseminator-of-chaos" (1990, 102).

9. The division of the glacier's three tongues among the *ukukus* goes back to the early 1960s. Sallnow reports from his fieldwork in 1984: "Four separate snowfields overhang the Sinakara valley. Formerly, only the two central ones were occupied by ukukus, the one on the left by those from the Paucartambo moiety and the one to the right by those from the Quispicanchis, but since 1980 the Canchis group has taken over one of the vacant glaciers" (1987, 227).

10. According to Salas Carreño (2006), the tradition of bringing chunks of ice to Cuzco's procession of Corpus Christi dates back to the late 1980s and 1990s, when the *ukuku* dancers of San Sebastián started accompanying the saint's image upon their return from Quyllurit'i.

11. In 2016 the Institute of Water Management under Cusco's regional government collected 10 tons of organic and 17 tons of inorganic waste at Sinak'ara after the pilgrimage (Salas Carreño 2021, 70).

12. In fact, Consejo de Naciones had already introduced its own ban on cutting ice from the glacier (Salas Carreño 2021).

13. Ceruti's account is unusual because the police are usually absent at the pilgrimage.

14. Apparently the Ministry of Culture has not been able to put any marks around the shrine's protected area, mainly because the local community opposes this demarcation (Salas Carreño 2020).

15. Stories about dead *ukukus* are difficult to verify. While I was told the stories cited in this chapter by people who claimed to have seen the dead bodies on the glacier, I was not provided information about the exact circumstances of the accidents. Considering their tragic consequences and the appeal they could have to existing stereotypes about Andean people, we should be cautious not to overestimate their numbers and frequency.

16. More specifically, Salas Carreño describes a certain type of foreign and Peruvian, but nonlocal, pilgrims as "New Age pilgrims."

REFERENCES

Abercrombie, Thomas. 1998. *Pathways of Memory and Power: Ethnography and History Among an Andean People*. Madison: University of Wisconsin Press.

Adger, Neil. 2006. "Vulnerability." *Global Environmental Change* 16: 268–281.

Adger, Neil, Jon Barnett, Katrina Brown, Nadine Marshall, and Karen O'Brien. 2012. "Cultural Dimensions of Climate Change Impacts and Adaptation." *Nature Climate Change* 3: 112–117.

Adger, Neil, Saleemun Huq, Katrina Brown, Declan Conway, and Mike Hulme. 2003. "Adaptation to Climate Change in the Developing World." *Progress in Development Studies* 3 (3): 179–195.

Aisher, Alex, and Vinita Damodaran. 2016. "Introduction. Human-Nature Interactions through Multispecies Lens." *Conservation and Society* 14 (4): 293–304.

Allen, J. Catherine. 1988. *The Hold Life Has: Coca and Cultural Identity in an Andean Community*. Washington, DC: Smithsonian Institution Press.

Allison, Elisabeth. 2015. "The Spiritual Significance of Glaciers in an Age of Climate Change." *WIREs Climate Change* 6 (5): 493–508.

Allouche, Jeremy. 2016. "The Birth and Spread of IWRM—A Case Study of Global Policy Diffusion and Translation." *Water Alternatives* 9 (3): 412–433.

Altamirano, Teófilo R. 2014. *Refugiados Ambientales: Cambio climático y migración forzada*. Lima: Fondo Editorial de Pontifica Universidad Católica del Perú.

Altamirano, Teófilo, and Erick Altamirano. 2019. *La nueva cucina peruana: En la era del cambio climático, la contaminación ambiental, las migraciones and la masculinización*. Lima: FEMIP (Federación de Instituciones de Peruanos en el mundo).

ANA (Autoridad Nacional del Agua). 2010. *Ley de Recursos Hídricos y su Reglamento*. Ley No 29338. Lima: Ministerio de Agricultura.

———. 2014. *Inventario de glaciares*. Lima: Autoridad Nacional de Agua. www.ana.gob.pe/media/981508/glaciares.pdf.

Andersen, Astrid O. 2016. "Infrastructures of Progress and Dispossession: Collective Responses to Shrinking Water Access among Farmers in Arequipa, Peru." *Focaal—Journal of Global and Historical Anthropology* 74: 28–41.

———. 2019. "Assembling Commons and Commodities: The Peruvian Water Law between Ideology and Materialization." *Water Alternatives* 12 (2): 470–487.

Andolina, Robert. 2012. "The Values of Water: Development Cultures and Indigenous Cultures in Highland Ecuador." *Latin American Research Review* 47 (2): 3–26.

Ansión, Juan. 1987. *Desde el rincón de los muertos: El pensamiento mítico en Ayacucho.* Lima: Gredes (Grupo de Estudios para el Desarrollo).

Arroya Aliaga, Jacinto, Natalie Schulz, and Pedro Gurmendi Párraga. 2012. "Impactos de las Actividades en el Nevado Huaytapallana." *Apuntes de Ciencia & Sociedad* 2 (1): 3–14.

Attala, Luci. 2019. *How Water Makes Us Human: Engagements with the Materiality of Water.* Cardiff: University of Wales Press.

Baer, Hans, and Merrill Singer. 2014. *The Anthropology of Climate Change: An Integrated Critical Perspective.* London: Routledge.

Bakker, Karen. 2007. "The 'Commons' versus the 'Commodity': Alter-globalization, Anti-privatization and the Human Right to Water in the Global South." *Antipode* 39 (3): 430–455.

———. 2012. "Water: Political, Biopolitical and Material." *Social Studies of Science* 42 (4): 616–624.

Barnes, Jessica, and Michael Dove, eds. 2015. *Climate Cultures: Anthropological Perspectives.* New Haven, CT: Yale University Press.

Barnes, Jessica, Michael Dove, Mianna Lahsen, Andrew Mathews, Pamela McElwee, Roderick McIntosh, Frances Moore, Jessica O'Reilley, Ben Orlove, Rajinda Puri, Harvey Weiss, and Karina Yager. 2013. "Contributions of Anthropology to the Study of Climate Change." *Nature Climate Change* 3: 541–544.

Barnett, T. P., J. C. Adam, and D. P. Lettenmaier. 2005. "Potential Impacts of a Warming Climate on Water Availability in Snow-Dominated Regions." *Nature* 438: 303–309.

Bastien, Joseph. 1978. *Mountain of the Condor: Metaphor and Ritual in an Andean Ayllu.* Prospect Heights, IL: Waveland.

———. 1985. "Qollohueraya-Andean Body Concepts: A Topographical-Hydraulic Model of Physiology." *American Anthropologist* 87 (3): 595–611.

Bauer, Adrian, and Mona Bhan. 2018. *Climate without Nature: A Critical Anthropology of the Anthropocene.* Cambridge: Cambridge University Press.

Beck, Ulrick. 2010. "Remapping Social Inequalities in an Age of Climate Change: For a Cosmopolitan Renewal of Sociology?" *Global Networks* 10 (2): 165–181.

Beltrán, María, and Esther Velásquez. 2017. "The Political Ecology of Water Metabolism: The Case of the Cobre las Cruces Copper Mine, Southern Spain." *Sustainable Science* 12: 333–343.

Beniston, Michael. 2003. "Climatic Change in Mountain Regions: A Review of Possible Impacts." *Climatic Change* 59 (1–2): 5–31.

Beresford, Melissa. 2020. "The Embedded Economics of Water: Insights from Economic Anthropology." *WIRE's Water* 7 (4): e1443.

Besom, Thomas. 2013. *Inka Human Sacrifice and Mountain Worship: Strategies for Empire Unification.* Albuquerque: University of New Mexico Press.

Biersack, Aletta. 2006. "Reimagining Political Ecology: Culture/Power/History/ Nature." In *Reimagining Political Ecology*, edited by Aletta Biersack and James B. Greenberg, 3–40. Durham, NC: Duke University Press.

Blaser, Mario. 2013. "Ontological Conflicts and the Stories of Peoples in Spite of Europe: Toward a Political Ontology." *Current Anthropology* 54 (5): 547–568.

———. 2016. "Is Another Cosmopolitics Possible?" *Cultural Anthropology* 31 (4): 545–570.

Boelens, Rudgerd. 2008. "Water Rights Arenas in the Andes: Upscaling Networks to Strengthen Local Water Control." *Water Alternatives* 1 (1): 48–65.

———. 2009. "The Politics of Disciplining Water Rights." *Development and Change* 40 (2): 307–333.

———. 2014. "Cultural Politics and the Hydrosocial Cycle: Water, Power and Identity in the Andean Highlands." *Geoforum* 57: 234–247.

———. 2015. *Water, Power and Identity: The Cultural Politics of Water in the Andes.* London: Routledge.

Bolin, Inge. 1998. *Rituals of Respect: The Secret of Survival in the High Peruvian Andes.* Austin. University of Texas Press.

———. 2009. "The Glaciers of the Andes Are Melting: Indigenous and Anthropological Knowledge Merge in Restoring Water Resources." In *Anthropology & Climate Change: From Encounters to Actions*, edited by Susan Crate and Mark Nuttall, 228–239. Walnut Creek, CA: Left Coast Press.

Brachetti, Angela. 2002. "Qoyllurrit'i: Una creencia andina bajo conceptos cristianos." *ANALES del Museo de América* (Madrid) 10: 85–112.

Bradley, Raymond, Mathias Vuille, Henry Diaz, and Walter Vergara. 2006. "Threats to Water Supplies in the Tropical Andes." *Science* 312 (5781): 1755–1756.

Brandshaug, Malene. 2019. "Water as More Than Commons or Commodity: Understanding Water Management Practices in Yanque, Peru." *Water Alternatives* 12 (2): 538–553.

———. 2021. "Water, Life, and Loss: Aguasociality and Environmental Change in the Peruvian Andes." *kritisk etnografi: Swedish Journal of Anthropology* 4 (2): 51–66.

Brügger, Adrian, Robert Tobias, and Fredy Monge-Rodríguez. 2021. "Public Perceptions of Climate Change in the Peruvian Andes." *Sustainability* 13 (3): 26–77.

Burman, Anders. 2017. "The Political Ontology of Climate Change: Moral Meteorology, Climate Justice, and the Coloniality of Reality ion the Bolivian Andes." *Journal of Political Ecology* 24 (1): 921–938.

Bury, Jeffrey, Bryan Mark, Jeffrey McKenzie, Adam French, Michel Baraer, Kuyung In Huh, Marco A. Zapata L., and Ricardo J. Gómez L. 2011. "Glacier Recession and Human Vulnerability in the Yanamarey Watershed of the Cordillera Blanca, Peru." *Climatic Change* 105 (1–2): 179–206.

Caine, Allison. 2021. "'Who Would Watch the Animals?': Gendered Knowledge and Expert Performance among Andean Pastoralists." *Culture, Agriculture, Food and Environment* 43 (1): 4–13.

Calder, Gideon. 2015. "Local Natures? Climate Change, Beliefs, Facts and Norms." *Climatic Change* 133 (3): 525–533.

Callison, Candis. 2014. *How Climate Change Comes to Matter: The Communal Life of Facts*. Durham, NC: Duke University Press.

Cammack, Richard, Teresa Atwood, Peter Campbell, Howard Parish, Anthony Smith, Frank Vella, and John Stirling. 2008. "Metabolism." In *Oxford Dictionary of Biochemistry and Molecular Biology*, 2nd ed. Oxford: Oxford University Press.

Campbell, Ben. 2008. "Environmental Cosmopolitans." *Nature and Culture* 3 (1): 9–24.

Carey, Mark. 2010. *In the Shadow of Melting Glaciers: Climate Change and Andean Society*. Oxford: Oxford University Press.

Carey, Mark, Olivia Molden, Mattias B. Rasmussen, M. Jackson, Anne Nolin, and Bryan Mark. 2017. "Impacts of Glacier Recession and Declining Meltwater on Mountain Societies." *Annals of the Association of American Geographers* 107 (2): 350–359.

Cebon, Peter, Urs Dahinden, Huw Davies, Dieter Imboden, and Carlo Jaerger. 1998. *Views from the Alps: Regional Perspectives on Climate Change*. Cambridge, MA: MIT Press.

Ceruti, María Constanza. 2007. "*Qoyllyr Riti*: Etnografía de un peregrinaje ritual de raíz incaíca por las altas montaños del Sur de Perú." *Scripta Ethnologica* 29, Bs. As: 9–35.

———. 2013. "Sacred Ice Melting Away: Lessons from the Impact of Climate Change on Andean Cultural Heritage." *Journal of Sustainability Education* 4: 1–9.

Chakrabarty, Dipesh. 2017. "The Politics of Climate Change Is More Than the Politics of Capitalism." *Theory, Culture & Society* 34 (2–3): 25–37.

Choy, Timothy, Lieba Faier, Michael Hathaway, Miyako Inoue, Shiho Satsuka, and Anna Tsing. 2009. "A New Form of Collaboration in Cultural Anthropology: Matsutake Worlds." *American Ethnologist* 36 (2): 380–403.

Climatelinks. 2017. "Climate Risk Profile: Peru." February 8. www.climatelinks.org/resources/climate-change-risk-profile-peru.

Colloredo-Mansfeld, Rudi. 2009. *Fighting Like a Community: Andean Civil Society in an Era of Indian Uprisings*. Chicago: University of Chicago Press.

Cometti, Geremia. 2020a. "A Cosmopolitical Ethnography of a Changing Climate among the Q'ero of the Peruvian Andes." *Anthropos* 115: 27–52,

———. 2020b. "El Antropoceno puesto a prueba en el campo: Cambio climático y crisis de las relaciones de reciprocidad entre los q'eros de los Andes peruanos." *Antípoda. Revista de Antropología y Arqueología* 38: 3–23.

Cook, David Noble. 1981. *Demographic Collapse: Indian Peru, 1520–1620*. Cambridge: Cambridge University Press.

———. 2007. *People of the Volcano: Andean Counterpoint in the Colca Valley of Peru*. Durham, NC: Duke University Press.

Cotler, Julio, and Ricardo Cuenca, eds. 2011. *Las desigualidades en el Perú: Balance críticos*. Lima: IEP (Instituto de Estudios Peruanos).

Coudrain, Anne, Bernard Francou, and Zbienew Kundzewicz. 2005. "Glacier Shrinkage in the Andes and Consequences for Water Resources." *Hydrological Sciences Journal* 50: 925–932.

Crate, Susan. 2011. "Climate and Culture: Anthropology in the Era of Contemporary Climate Change." *Annual Review of Anthropology* 40: 175–194.

Crate, Susan, and Mark Nuttall, eds. 2009. *Anthropology & Climate Change: From Encounter to Actions*. Walnut Creek, CA: Left Coast Press.

——, eds. 2016. *Anthropology & Climate Change: From Encounter to Actions*. 2nd ed. New York: Routledge.

Crona, Beatrice, Amber Wutich, Alexandra Brewis, and Meredith Gartin. 2013. "Perceptions of Climate Change: Linking Local and Global Perceptions through a Cultural Knowledge Approach." *Climatic Change* 119 (2): 519–531.

Cruikshank, Julie. 2005. *Do Glaciers Listen? Local Knowledge, Colonial Encounters & Social Imagination*. Vancouver: University of British Columbia Press.

de la Cadena, Marisol. 2000. *Indigenous Mestizos: The Politics of Race and Culture in Cuzco, 1919–1991*. Durham, NC: Duke University Press.

——. 2010. "Indigenous 'Cosmopolitics' in the Andes: Conceptual Reflections beyond 'Politics.'" *Cultural Anthropology* 25 (2): 334–370.

——. 2015. *Earth beings: Ecologies of Practice across Andean Worlds*. Durham, NC: Duke University Press.

de la Cadena, Marisol, and Mario Blaser, eds. 2018. *A World of Many Worlds*. Durham, NC: Duke University Press.

del Castillo Pinto, Laureano. 2011. "Ley de recursos hídricos: necesaria pero no suficiente." *Debate Agrario* 45: 91–118.

Damonte, Gerardo, and Rutgerd Boelens. 2019. "Hydrosocial Territories, Agroexport and Water Scarcity: Capitalist Territorial Transformations and Water Governance in Peru's Coastal Valley." *Water International* 44 (2): 206–223.

Derman, Bill, and Anne Ferguson. 2003. "Value of Water: Political Ecology and Water Reform in Southern Africa." *Human Organization* 62 (3): 277–288.

Descolá, Philippe. 1996. "Constructing Nature: Symbolic Ecology and Social Practice." In *Nature and Society: Anthropological Perspectives*, edited by Philippe Descolá and Gísli Pálsson, 82–102. London: Routledge.

——. 2013. *Beyond Nature and Culture*. Chicago: University of Chicago Press.

Diaz, Henry, Martin Grosjean, and Lisa Graumlich. 2003. "Climate Variability and Change in High Elevation Regions: Past, Present and Future." *Climatic Change* 59 (1–2): 1–4.

Drenkhan, Fabian, Mark Carey, Christian Hüggel, Jochen Seidel, and María Teresa Oré. 2015. "The Changing Water Cycle: Climatic and Socioeconomic Drivers of Water-Related Changes in the Andes of Peru." *WIREs Water* 2 (6): 715–733.

Dwyer, Peter D. 1985. Review of *Pigs for the Ancestors: Ritual in the Ecology of a New Guinea People; A New Enlarged Edition*. *Oceania* 56: 151–154.

Emmett, Robert, and Thomas Lekan, eds. 2016. *Whose Anthropocene? Revisiting Dipesh Chakrabarty's "Four Theses"*. RCC Perspectives/Transformations in Environment and Society. Berlin: Federal Ministry of Education and Research.

ENCC (Estrategia Nacional ante el Cambio Climático). 2015. *Estrategia nacional ante el cambio climático*. Lima: Ministerio del Ambiente.

Escobar, Arturo. 1999. "After Nature: Steps to an Antiessentialist Political Ecology." *Current Anthropology* 40 (1): 1–16.

Farhana, Sultana, and Alex Loftus. 2015. "The Human Right to Water: Critiques and Conditions of Possibility." *WIREs Water* 2 (2): 97–105.

Fine-Dare, Kathleen S. 2019. *Urban Mountain Beings: History, Indigeneity, and Geographies of Time in Quito, Ecuador.* Lanham, MD: Rowman & Littlefield.

Flores Lizana, Carlos. 1997. *El Taytacha Qoyllur Rit'i: Teología india hecha por comuneros y mestizos quechuas.* Cuzco: Instituto de Pastoral Andina.

Folke, Carl. 2006. "Resilience: The Emergence of a Perspective for Social-Ecological Systems Analyses." *Global Environmental Change* 16: 253–267.

Funtowicz, Silvio, and Jerome Ravetz. 1993. "Science for Post-normal Age." *Future* 25 (7): 735–755.

Gagné, Karine, Mattias Rasmussen, and Benjamin Orlove. 2014. "Glaciers and Society: Attributions, Perceptions, and Valuations." *WIREs Climate Change* 5: 793–808.

Gálaz, Victor. 2004. "Stealing from the Poor? Game Theory and the Politics of Water Markets in Chile." *Environmental Politics* 13(2): 414–437.

García, María Elena. 2005. *Making Indigenous Citizens. Identity, Development, and Multicultural Activism in Peru.* Stanford, CA: Stanford University Press.

Gareis, Iris. 2019. "Andean Gods and Catholic Saints: Indigenous and Catholic Intercultural Encounters." In *The Andean World*, edited by Linda Seligmann and Kathleen Fine-Dare, 266–279. London: Routledge.

Gelles, Paul. 1994. "Channels of Power, Fields of Contention: The Politics of Irrigation and Social Organization of Water Control in an Andean Peasant Community." In *Irrigation at High Altitudes: The Social Organization of Water Control in the Andes*, edited by William Mitchell and David Guillet, 233–273. Washington, DC: Society for Latin American Anthropology and the American Anthropological Association.

———. 2000. *Water and Power in Highland Peru: The Cultural Politics of Irrigation and Development.* New Brunswick, NJ: Rutgers University Press.

———. 2005. "Transformaciónes en una comunidad andina transnacional." In *El Quinto Suyo: Transnacionalidad y formaciones diaspóricas en la migración peruana*, edited by Ulla Berg and Karsten Paerregaard, 69–96. Lima: Instituto de Estudios Peruanos.

Golte, Jürgen, and Doris León Gabriel. 2014. *Alasitas: Discursos, prácticas y símbolos de un "Liberalismo Aymara Altiplánico" entre la población de origen migrante en Lima.* Lima: IEP (Instituto de Estudios Peruanos).

Gómez, Guillermo C., and Roy G. Santos. 2012. "Riesgos de escasez de agua en la ciudad de Huancayo al año 2030." *Apuntos Ciencias Sociales* 2 (1): 15–26.

Gootenberg, Peter. 1991. "Population and Ethnicity in Early Republican Peru: Some Revisions." *Latin American Research Review* 26 (3): 109–157.

Gorriti, Gustavo. 1999. *The Shining Path: A History of the Millenarian Water in Peru.* Chapel Hill: University of North Carolina Press.

Gose, Peter. 1986. "Sacrifice and the Commodity Form in the Andes." *Man (Royal Anthropological Institute of Britain and Ireland)* 21 (2): 296–310

———. 1994. *Deathly Waters and Hungry Mountains: Agrarian Ritual and Class Formation in an Andean Town*. Toronto: University of Toronto Press.

———. 2008. *Invaders as Ancestors: On the Intercultural Making and Unmaking of Spanish Colonialism in the Andes*. Toronto: University of Toronto Press.

———. 2018. "The Semi-social Mountain: Metapersonhood and Political Ontology in the Andes." *HAU: Journal of Ethnographic Theory* 8 (3): 488–505.

———. 2019. "The Andean Circulatory Cosmos." In *The Andean World*, edited by Linda Seligmann and Kathleen Fine-Dare, 115–127. London: Routledge.

Gow, David. 1974. "Taytacha Qoyllur Rit'i: Rocas y bailarines; Creencia y continuidad." *Allpanchis Phuturinqa* 7 (1): 49–100.

Graeter, Stefanie. 2017. "To Revive an Abundant Life: Catholic Science and Neoextractivist Politics in Peru's Mantary Valley." *Cultural Anthropology* 32 (1): 117–148.

Green, D., and G. Raygorodetskty. 2010. "Indigenous Knowledge of a Changing Climate." *Climatic Change* 100 (2): 239–242.

Greene, Shane. 2009. *Customizing Indigeneity: Paths to a Visionary Politics in Peru*. Stanford, CA: Stanford University Press.

Gremaud, Ann-Sofie. 2014. "Power and Purity: Nature as Resource in a Troubled Society." *Environmental Humanities* 5: 77–100.

Greschke, Heike. 2015. "The Social Facts of Climate Change: An Ethnographic Approach." In *Grounding Global Climate Change: Contributions from the Social and Cultural Sciences*, edited by Heike Greschke and Julia Tischler, 121–138. New York: Springer Press.

Greschke, Heike, and Julia Tischler, eds. 2015. *Grounding Global Climate Change: Contributions from the Social and Cultural Sciences*. New York: Springer Press.

Groenfeldt, David. 2013. *Water Ethics: A Values Approach to Solving the Water Crisis*. London, England: Routledge.

Guillet, David. 1992. *Covering Ground: Communal Water Management and the State in the Peruvian Highlands*. Ann Arbor: University of Michigan Press.

Gustavsson, Maria-Therese. 2016. "Private Conflict Regulation and the Influence of Indigenous Peasants over Natural Resources." *Latin American Research Review* 51 (2): 86–106.

Haller, Andreas, and Hildegardo Córdova-Aguilar. 2018. "Urbanization and the Advent of Regional Conservation: Huancayo and the Cordillera Huaytapallana, Peru." *Management & Policy Issues* 10 (2): 59–63.

Harris, Olivia. 1982. "The Dead and the Devils among the Bolivian Laymi." In *Death and the Regeneration of Life*, edited by Maurice Bloch and Jonathan Parry, 45–73. Cambridge: Cambridge University Press.

Hastrup, Kirsten. 2013. "Anthropological Contributions to the Study of Climate: Past, Present, Future." *WIREs Climate Change* 4 (4): 269–281.

———. 2016. "Climate Knowledge: Assemblage, Anticipation, Action." In *Anthropology and Climate Change: From Encounters to Action*, 2nd ed., edited by Susan Crate and Mark Nuttall, 35–57. New York: Routledge.

Hastrup, Kirsten, and Karen F. Olwig, eds. 2012. *Climate Change and Human Mobility: Challenges to the Social Sciences*. Cambridge: Cambridge University Press.

Hastrup, Kirsten, and Martin Skydstrup, eds. 2012. *The Social Life of Climate Change Models: Anticipating Nature*. London: Routledge.

Helmreich, Stefan. 2011. "Nature/Culture/Seawater." *American Anthropologist* 113 (1): 132–144.

Huayhua, Margarita. 2014. "Racism and Social Interaction in a Southern Peruvian Combi." *Ethnic and Racial Studies* 37 (13): 2399–2417.

Hubert, Henri, and Marcel Mauss. 1964. *Sacrifice: Its Nature and Function*. Chicago: University of Chicago Press.

Hulme, Mike. 2009. *Why We Disagree about Climate: Understanding Controversy, Inaction and Opportunity*. Cambridge: Cambridge University Press.

———. 2011. "Reducing the Future to Climate: A Story of Climate Determinism and Reductionism." *Osiris* 26: 245–266.

INAIGEM (Instituto Nacional de Investigación en Glaciares y Ecosistemas de Montaña). 2017. *Informe de la situación de los glaciares y ecosistemas de montaña en el Perú*. Lima: INIAGEM, Dirección de Información y Gestión del Conocimiento.

INEI (Instituto de Estadística y Informática–Peru). 2017. *Censo nacionales 2017, XII de población*. Lima: INEI-CPV 2017.

Ingold, Tim. 2007. "Earth, Sky, Wind and Weather." *Journal of the Royal Anthropological Institute*, n.s., 16, supp. 1: S19–S38.

———. 2012. "Toward an Ecology of Materials." *Annual Review of Anthropology*, no. 41: 427–442.

Isbell, Billie Jean. 1978. *To Defend Ourselves: Ecology and Ritual in an Andean Village*. Prospect Heights, IL: Waveland.

Isch, Edgar L., Rutgerd Boelens, and Francisco Peña, eds. 2012. *Agua, injusticia y conflictos*. Lima: Justicia Hídrica, Centro de Studios Regionales Andinos, Fondo Editorial PUCP, IEP (Instituto de Estudios Peruanos).

Jasanoff, Sheila. 2010. "A New Climate for Society." *Theory, Culture & Society* 27 (2–3): 233–253.

Jurt, Christine, María Dulce Burga, Luís Vicuna, Christian Hüggel, and Ben Orlove. 2015. "Local Perceptions in Climate Change Debates: Insights from Case Studies in the Alps and the Andes." *Climatic Change* 133 (3): 511–523.

Kirksey, S. Eben, and Stefan Helmreich. 2010. "The Emergence of Multispecies Ethnography." *Cultural Anthropology* 25 (4): 545–576.

Krauss, Werner, and Hans von Storch. 2012. "Post-Normal Practices between Regional Climate Services and Local Knowledge." *Nature and Culture* 7 (2): 213–230.

Kubler, George. 1952. *The Indian Caste of Peru, 1795–1940: A Population Study Based upon Tax Records and Census Reports*. Washington, DC: Smithsonian Institute.

Kundzewiz, Z. W., Luís Mata, Nigel Arnell, P. Döll, Blanca Jimenez, Kathleen Miller, Taikan Oki, Zekai Sen, and I. Shiklomanov. 2008. "The Implications of Projected Climate Change for Freshwater Resources and Their Management." *Hydrological Sciences Journal* 53 (1): 3–10.

Lambeck, Michael. 2011. "Catching the Local." *Anthropological Theory* 11 (2): 197–221.

Lanata, Xavier Ricard. 2007. *Ladrones de sombra: El universo religioso de los pastores de Ausangate*. Lima: IFEA/CBC.

Latour, Bruno. 2011. "From Multiculturalism to Multinaturalism: What Rules of Methods for the New School of Social-Scientific Experiments?" *Nature and Culture* 6 (1): 1–7.

———. 2013. *An Inquiry into Modes of Existence: An Anthropology of the Moderns*. Cambridge, MA: Harvard University Press.

Lennox, Erin, and John Gowdy. 2014. "Ecosystem Governance in a Highland Village in Peru: Facing the Challenges of Globalization and Climate Change." *Ecosystem Services* 10: 155–163.

Li, Fabiana. 2015. *Unearthing Conflict: Corporate Mining, Activism, and Expertise in Peru*. Durham, NC: Duke University Press.

Linton, Jamie, and Jessica Budds. 2014. "The Hydrosocial Cycle: Defining and Mobilizing a Relational-dialectical Approach to Water." *Geoforum* 57: 170–180.

López-Moreno, J. I., S. Fontaneda, J. Bazo, J. Revuelto, C. Azorin-Molina, B. Valero-Garcés, E. Morán-Tejeda, S. M. Vicente-Serrano, R. Zubieta, and J. Alejo-Cochachín. 2014. "Recent Glacier Retreat and Climate Trends in Cordillera Huaytapallana, Peru." *Global and Planetary Change* 112: 1–11.

Lowell, Nadia. 1998. Introduction to *Locality and Belonging*, edited by Nadia Lowell, 1–24. London: Routledge.

Lynch, Barbara D. 2012. "Vulnerabilities, Competition and Rights in a Context of Climate Change toward Equitable Water Governance in Peru's Rio Santa Valley." *Global Environmental Change* 22 (2): 364–373.

———. 2019. "Water and Power in the Peruvian Andes." In *The Andean World*, edited by Linda Seligmann and Kathleen Fine-Dare, 44–59. London: Routledge.

Lynch, Nicolás. 2014. *Cholofícación, república y democracia: El destino negado del Perú*. Lima: OtraMirada.

MacCormack, Sabine. 1991. *Religion in the Andes: Vision and Imagination in Early Colonial Peru*. Princeton, NJ: Princeton University Press.

Madrid-López, Cristina, and Mario Giampietro. 2015. "The Water Metabolism of Socio-ecological Systems: Reflections and a Conceptual Framework." *Journal of Industrial Ecology* 19 (5): 853–865.

Marcus, George. 1998. *Ethnography through Thick & Thin*. Princeton, NJ: Princeton University Press.

Mariátegui, José Carlos. 1974. *Seven Interpretive Essays on Peruvian Reality*. Austin: University of Texas Press.

Mark, Bryan G., Adam French, Michel Baraer, Mark Carey, Jeffrey Bury, Kenneth Young, Molly Polk, Oliver Wigmore, Pablo Lagos, Ryan Crumley, Jeffrey McKenzie, and Laura Lautz. 2017. "Glacier Loss and Hydro-social Risks in the Peruvian Andes." *Global and Planetary Change* 159: 61–76.

Marx, Karl. 1992. *Economics and Philosophical Manuscripts: From Early Writings*. New York: Penguin Books.

Mathur, Nayanika. 2015. "It's a Conspiracy Theory *and* Climate Change: Of Beastly Encounters and Cervine Disappearances in Himalayan India." *HAU: Journal of Ethnographic Theory* 5 (1): 87–111.

Mayer, Enrique. 2009. *Ugly Stories of the Peruvian Agrarian Reform*. Durham, NC: Duke University Press.

Mendoza, Zoila. 2010. "La fuerza de los caminos sonoros: Caminate y música de Qoyllur Rit'i." *Anthropologica* 28: 15–38.

Milan, Andrea, and Raul Ho. 2014. "Livelihood and Migration Patterns at Different Altitudes in the Central Highlands of Peru." *Climate and Development* 6 (1): 69–76.

Ministerio de Ambiente. 2015. *Portal de Cambio Climático*. http://cambioclimatico .minam.gob.pe/manejo-de-la-tierra-y-el-agua/manejo-del-agua/impactos-del-cc -sobre-el-agua.

Mitchell, William. 2006. *Voices from the Global Margin: Confronting Poverty and Inventing New Lives in the Andes*. Austin: University of Texas Press.

Mitchell, William, and David Guillet, eds. 1994. *Irrigation at High Altitudes: The Social Organization of Water Control in the Andes*. Washington, DC: AAA.

Molinié, Antoinette. 2019. "Indian Identity and Indigenous Revitalization Movements." In *The Andean World*, edited by Linda Seligmann and Kathleen Fine-Dare, 373–388. London: Routledge.

Moulton, Holly, Mark Carey, Christian Huggel, and Alina Motschmann. 2021. "Narratives of Ice Loss: New Approaches to Shrinking Glaciers and Climate Change Adaptation." *Geoforum* 125: 47–56.

Nash, June. 1979. *We Eat the Mines and the Mines Eat Us: Dependency and Exploitation in Bolivian Tin Mines*. New York: Columbia University Press.

Nelson, Velvet. 2016. "Peru's Image as a Culinary Destination." *Journal of Cultural Geography* 33 (2): 208–228:

Ødegaard, Cecilie Vindal. 2011. "Sources of Danger and Prosperity in the Peruvian Andes: Mobility in a Powerful Landscape." *Journal of the Royal Anthropological Institute* 17 (2): 339–355.

———. 2020. *Mobility, Markets, and Indigenous Socialities: Contemporary Migration in the Peruvian Andes*. London: Routledge

Ødegaard, Cecilie Vindal, and Juan Rivera Andía, eds. 2019. *Indigenous Life Projects and Extractivism: Ethnographies from South America*. Cham, Switzerland: Palgrave Macmillan.

Olivas Weston, Marcela. 1999. *Peregrinaciones en el Perú: Antiguas rutas devocionales*. Lima: Universidad de San Martín de Porres.

Oliver-Smith, Anthony. 2016. "The Concepts of Adaptation, Vulnerability, and Resilience in the Anthropology of Climate Change: Considering the Case of Displacement and Migration." In *Anthropology and Climate Change: From Encounters to Action*, 2nd ed., edited by Susan Crate and Mark Nuttall, 68–85. New York: Routledge.

Oliver-Smith, Anthony, and Xiaomeng Shen, eds. 2009. *Linking Environmental Change, Migration and Social Vulnerability*. Bonn: United Nations University.

Oré, María Teresa. 2005. *Agua bien y usos privados*. Lima: Fondo Editorial de PUCP.

Oré, María Teresa, Laureano del Castillo, Saskia Van Orsel, and Jeroen Vos. 2009. *El agua, ante nuevos desafíos: Actores e iniciativas en Ecuador, Perú y Bolivia*. Lima: Oxfam/IEP.

Orlove, Ben, and Steve Caton. 2010. "Water Sustainability: Anthropological Approaches and Prospects." *Annual Review of Anthropology* 39 (1): 401–415.

Orlove, Ben, and David Guillet. 1985. "Theoretical and Methodological Considerations of the Study of Mountain Peoples." *Mountain Research and Development* 5 (1): 3–18.

Orlove, Ben, Ellen Wiegandt, and Brian H. Luckman, eds. 2008. *Darkening Peaks Glacier Retreat, Science, and Society*. Berkeley: University of California Press.

Paerregaard, Karsten. 1987a. "Death Rituals and Symbols in the Andes." *Folk* 29: 23–42.

———. 1987b. *Nuevas organizaciones en comunidades campesinas: El caso de Usibamba y Chaquicocha*. Lima: Pontificia Universidad Católica del Perú.

———. 1989. "Exchanging with Nature: *T'inka* in an Andean Village." *Folk* 31: 53–73.

———. 1992. "Complementarity and Duality: Ecological, Social and Ritual Oppositions between Agriculturalists and Herders in an Andean Village." *Ethnology* 31(1): 15–26.

———. 1993. "Mas allá del dinero: Trueque y economía categorial en un distrito en el valle de Colca." *Antropológica* (Universidad Católica del Perú) 11: 211–251.

———. 1994a. "Why Fight over Water? Power, Conflict and Irrigation in an Andean Village." In *Irrigation at High Altitudes: The Social Organization of Water Control Systems in the Andes*, edited by David Guillet and William Mitchell, 189–202. Washington, DC: Society for Latin American Anthropology.

———. 1994b. "Conversion, Migration, and Social Identity: The Spread of Protestantism in the Peruvian Andes." *Ethnos* 59 (3–4): 168–186.

———. 1997a. *Linking Separate Worlds: Urban Migrants and Rural Lives in Peru*. Oxford: Berg.

———. 1997b. "Imagining a Place in the Andes: In the Borderland of Analyzed, Invented, and Lived Culture." In *Siting Culture: The Shifting Anthropological Object*, edited by Kirsten Hastrup and Karen F. Olwig, 39–59. Oxford: Routledge.

———. 1998. "The Dark Side of the Moon: Conceptual and Methodological Problems of Studying Urban Migrants and Their Native Village." *American Anthropologist* 100 (2): 397–408.

———. 2000. "Procesos migratorios y estrategias complementarias en la sierra peruana." *European Review of Latin American and Caribbean Studies* 69: 69–80.

———. 2002. "The Vicissitudes of Politics and the Resilience of the Peasantry: The Contestation and Reconfiguration of Political Space in the Peruvian Andes." In *In the Name of the Poor: Contesting Political Space for Poverty Reduction*, edited by Neil Webster and L. Engberg-Pedersen, 52–77. London: Zed Books.

———. 2005. "Inside the Hispanic Melting Pot: Negotiating National and Multicultural Identities among Peruvians in the United States." *Journal of Latino Studies* 3(2): 76–96.

———. 2008a. *Peruvians Dispersed: A Global Ethnography of Migration*. Lanham, MD: Lexington Books.

———. 2008b. "In the Footsteps of the Lord of the Miracles: The Expatriation of Religious Symbols in the Peruvian Diaspora." *Journal of Ethnic and Migration Studies* 34 (7): 1073–1089

———. 2010a. "The Show Must Go On: The Role of Fiesta in Andean Transnational Migration." *Latin American Perspectives* 37 (5): 50–66.

———. 2010b. "Globalizing Andean Society: Migration and Change in Peru's Peasant Communities." In *Vicos Experience: New Perspectives on Rural Development in Peru*, edited by Tom Greaves and Ralph Bolton, 95–213. San Francisco: AltaMira Press.

———. 2011. "Transnational Crossfire: Local, National and Global Conflicts in Peruvian Migration." In *Local Battles, Global Stakes: The Globalization of Local Conflicts and the Localization of Global Interests*, edited by Ton Salman and Marjo de Theije, 155–174. Amsterdam: VU University Press.

———. 2012a. "Commodifying Intimacy: Women, Work and Mobility in Peruvian Migration." *Journal of Latin American and Caribbean Anthropology* 17 (3): 493–511.

———. 2012b. "The Grass of Wrath: US Labor Migration and Poverty Alleviation in the Peruvian Andes." In *The Byways of the Poor: Organizing Practices and Economic Control among Rural Poor in the Third World*, edited by Karsten Paerregaard and Neil Webster, 227–250. Copenhagen: Museum Tusculanum.

———. 2013a. "Bare Rocks and Fallen Angels: Environmental Change, Climate Perceptions and Ritual Practice in the Peruvian Andes." *Religions* 4 (2): 290–305.

———. 2013b. "Governing Water in the Andean Community of Cabanaconde: From Resistance to Opposition and to Cooperation (and Back Again?)." *Mountain Research and Development* 33 (3): 207–214.

———. 2014a. "Movements, Moments, and Moods: Generational Unity and Strife in Peruvian Migration." *Ethnic and Racial Studies* 37 (11): 2129–2147.

———. 2014b. "Broken Cosmologies: Climate, Water and State in the Peruvian Andes." In *Anthropology and Nature*, edited by Kirsten Hastrup, 196–210. London: Routledge.

———. 2015a. *Return to Sender: The Moral Economy of Remittances in Peruvian Migration*. Berkeley: California University Press.

———. 2015b. "The Resilience of Migrant Money: How Gender, Generation and Class Shape Family Remittances in Peruvian Migration." *Global Networks* 15 (4): 503–518.

———. 2016. "Making Sense of Climate Change: Global Impacts, Local Responses and Anthropogenic Dilemmas in the Peruvian Andes." In *Anthropology and Climate Change: From Encounters to Action*, 2nd ed., edited by Susan Crate and Mark Nuttall, 250–260. New York: Routledge.

———. 2017. "*Ayni* Unbound: Cooperation, Inequality and Migration in the Peruvian Andes." *Journal of Latin American and Caribbean Anthropology* 22 (3): 459–474.

———. 2018a. "The Climate-Development Nexus: Using Climate Voices to Prepare Adaptation Initiatives in the Peruvian Andes." *Climate and Development* 10 (4): 360–368.

———. 2018b. "Power as/in/of Water: Revisiting the Hydrological Cycle in the Peruvian Andes." *WIRE's Water* 5 (2): 1–11.

———. 2018c. "Capitalizing on Migration: The Role of Weak and Strong Ties among Peruvian Entrepreneurs in the US, Spain, and Chile." *Migration Studies* 6 (1): 79–98.

———. 2019a. "Liquid Accountability: Water as a Common, a Public and a Private Good in the Peruvian Andes." *Water Alternatives* 12 (3): 488–502.

———. 2019b. "El Apu llora: Cómo el cambio climático y el derretimiento glacial transforman la ofrenda andina." In *Montañas y paisajes sagrados: Mundos religiosos, cambio climático y el derretimiento glacial*, edited by Robert Albro, 101–131. Lima: Fondo Editorial Universidad Antonio Ruiz de Montoya.

———. 2019c. "Grasping the Fear: How Zenophobia Intersects with Climatephobia and Robotphobia and How Their Co-production Creates Feelings of Abandonment, Self-Pity and Destruction." *Migration Letters* 16 (4): 647–652.

———. 2019d. "Transnational Circuits: Migration, Money and Might in Peru's Andean Communities." In *The Andean World*, edited by Linda Seligmann and Kathleen Fine-Dare, 602–616. London: Routledge.

———. 2020a. "Communicating the Inevitable: Climate Awareness, Climate Discord and Climate Research in Andean Communities." *Environmental Communication* 14 (1): 112–125.

———. 2020b. "Searching for the New Human: Glacier Melt, Anthropogenic Change and Self-Reflection in Andean Pilgrimage." *HAU: Journal of Ethnographic Theory* 10 (3): 844–859.

———. 2021a. "Recasting the Sacred: Offering Ceremonies, Glacier Melt, and Climate Change in the Peruvian Andes." In *Understanding Climate Change through Religious Lifeworlds*, edited by David Haberman, 261–283. Bloomington: Indiana University Press.

———. 2021b. "Lubricating Water Metabolism: How Mountain Offerings Contribute to Water Sustainability in the Peruvian Andes." *kritisk etnografi: Swedish Journal of Anthropology* 4 (2): 83–98.

———. 2023. "Climing the Andes: Vertical Complementarity, Trans-Human Reciprocity, and Climate Change in the Peruvian Highlands." In *Storying Multipolar Climes of the Himalaya, Andes and Arctic*, edited by Dan Smyer Yü and Jelle J. P. Wouters, 52–68. London: Routledge.

Paerregaard, Karsten, and Astrid O. Andersen. 2019. "Moving Beyond the Commodity-Common Dichotomy: The Social-Political Complexity of Andean Water Governance." In special issue, *Water Alternatives* 12 (3): 459–469.

Paerregaard, Karsten, Astrid B. Stensrud, and Astrid O. Andersen. 2016. "Water Citizenship: Negotiating Water Rights and Contesting Water Culture in the Peruvian Andes." *Latin American Research Review* 51 (1): 198–217.

Paerregaard, Karsten, Susann Ullberg, and Malene Brandshaug. 2020. "Smooth Flow? Hydrosocial Communities, Water Governance, and Infrastructural Discord in the Peru's Southern Highlands." *Water International* 45 (3): 169–188.

Patterson, Thomas. 2009. *Karl Marx, Anthropologist*. Oxford: Berg.

Pérez Gálvez, Jesús Claudio, Tomás López-Guzmán, Franklin Cordova Buiza, and Miguel Jesús Medina-Viruel. 2017. "Gastronomy as an Element of Attraction in a Tourist Destination: The Case of Lima, Peru." *Journal of Ethnic Foods* 4 (4): 354–261.

Peru Reports. 2018. *Venezuelan Migrants in Peru Face Constant Challenges*. March 6. https://perureports.com/venezuelan-migrants-peru/6952/.

Poole, Deborah. 1988. "Entre el milagro y la mercancía: Qoyllur Rit'i." *Márgenes* 2 (4): 101–119.

———. 1990. "Accommodation and Resistance in Andean Ritual Dance." *Drama Review* 34 (2): 98–126.

Postigo, Julio. 2014. "Perception and Resilience of Andean Populations Facing Climate Change." *Journal of Ethnobiology* 43 (3): 383–400.

Poupeau, Franck, and Claudia González, eds. 2010. *Modelos de gestión del agua en los Andes*. Lima/La Paz: IFEA (Instituto Francés de Estudios Andinos/PIEB (Programa de Investigación Estratégica de Bolivia).

Ramírez, Juan Andrés. 1969. "La novena al Señor de Qoyllur Rit'i." *Allpanchis Phuturinqa* 1 (1): 61–87.

Ramirez, Susan Elisabeth. 1996. *The World Upside Down: Cross-Cultural Contact and Conflict in Sixteenth-Century Peru*. Stanford, CA: Stanford University Press.

Rangecroft, Sally, Stephan Harrison, Karen Anderson, John Magrath, Ana Paola Castel, and Paula Pacheco. 2013. "Climate Change and Water Resources in Arid Mountains: An Example from the Bolivian Andes." *Ambio* 42 (7): 852–863.

Rappaport, Roy A. 1968. *Pigs for the Ancestors*. New Haven, CT: Yale University Press.

Rasmussen, Mattias B. 2015. *Andean Waterways: Resource Politics in Highland Peru*. Seattle: University of Washington Press.

———. 2016a. "Water Futures: Contention in the Construction of Productive Infrastructures in the Peruvian Andes." *Anthropologica* 58 (2): 211–226.

———. 2016b "Reclaiming the Lake: Citizenship and Environment-as-Common-Property in Highland Peru." *Focaal—Journal of Global and Historical Anthropology* 74: 13–27.

———. 2016c. "Unsettling Times: Living with the Changing Horizons of the Peruvian Andes." *Latin American Perspectives* 43 (4): 73–86.

———. 2017. "Tactics of the Governed: Figures of Abandonment in Andean Peru." *Journal of Latin American Studies* 49 (2): 327–353.

Reinhard, Johan, and Maria Constanza Ceruti. 2010. *Inca Rituals and Sacred Mountains: A Study of the World's Highest Archaeological Sites*. Cotsen Institute of Archaeology, University of California Los Angeles (UCLA).

Roa-García, María Cecilia. 2014. "Equity, Efficiency and Sustainability in Water Allocation in the Andes." *Water Alternatives* 7 (2): 298–319.

Roa-García, María Cecilia, Patricia Urteaga-Crovetto, and Rocío Bustamante-Zenteno. 2015. "Water Laws in the Andes: A Promising Precedent for Challenging Neoliberalism." *Geoforum* 64: 270–280.

Rodríguez-Labajos, Beatrix, and Joan Martínez-Alier. 2015. "Political Ecology of Water Conflicts." *WIREs Water* 2 (5): 537–558.

Roncoli, Carla, Todd Crane, and Ben Orlove. 2009. "Fielding Climate Change in Cultural Anthropology." In *Anthropology & Climate Change: From Encounter to Actions*, edited by Susan Crate and Mark Nuttall, 116–137. Walnut Creek, CA: Left Coast Press.

Salas Carreño, Guillermo. 2006. "Diferenciación social y discursos públicos sobre la peregrinación de Quyllurit'i." In *Mirando la esfera pública desde la cultura en el Perú*, edited by Gisela Cánepa and María Ulfe, 243–288. Lima: Concytec.

———. 2014. "The Glacier, the Rock, the Image: Emotional Experience and Semiotic Diversity at the Quyllorit'i Pilgrimage." *Signs and Society* 2 (S1): 188–214.

———. 2017. "Mining and the Living Materiality of Mountains in Andean Societies." *Journal of Material Culture* 22 (2): 133–150.

———. 2019. *Lugares parientes: Comida, cohabitación y mundos andinos.* Lima: Pontificia Universidad Católica del Perú.

———. 2020. "Indexicality and the Indigenization of Politics: Dancers-Pilgrims Protesting Mining Concessions in the Andes." *Journal of Latin American and Caribbean Anthropology* 25 (1): 7–27.

———. 2021. "Climate Change, Moral Meteorology, and Local Measures at Quyllurit'i, a High Andean Shrine." In *Understanding Climate Change through Religious Lifeworlds*, edited by David Haberman, 44–76. Bloomington: Indiana University Press.

Salas Carreño, Guillermo, and Alejandro Diez Hurtado. 2018. "Estado, concesiones mineras y comuneros: Los múltiples conflictos alrededor de la minería en las inmediaciones del santuario de Qoyllurit'i (Cusco, Perú)." *Colombia Internacional* 93: 65–91.

Sallnow, Michael. 1974. "La peregrinación andina." *Allpanchis* 7: 101–142.

———. 1987. *Pilgrims of the Andes: Regional Cults in Cusco.* Washington, DC: Smithsonian Institute Press.

Salomon, Frank. 2018. *At the Mountains' Altar: Anthropology of Religion in an Andean Community.* London: Routledge.

Sayre, Nathan. 2012. "The Politics of the Anthropogenic." *Annual Review of Anthropology* 41: 57–70.

Schnegg, Michael, Coral Iris O'Brian, and Inga Janina Sievert. 2021. "It's Our Fault: A Global Comparison of Different Ways of Explaining Climate Change." *Human Ecology* 49: 327–339.

Seligmann, Linda. 1995. *Between Reform & Revolution: Political Struggles in the Peruvian Andes, 1969–1991.* Stanford, CA: Stanford University Press.

Shapero, Joshua. 2017. "Possessive Places: Spatial Routines and Glacier Oracles in Peru's Cordillera Blanca." *Ethnos* 84 (4): 615–641.

Sherbondy, Jeanette. 1982. "El regadío, los lagos y los mitos de origin." *Allpanchis* 20: 3–32.

———. 1994. "Water and Power: The Role of Irrigation Districts in the Transition from Inca to Spanish Cuzco." In *Irrigation at High Altitudes: The Social Organization of Water Control in the Andes*, edited by William Mitchell and David Guillet, 69–97. Washington, DC: AAA.

Starn, Orin. 1999. *Night Watch: The Politics of Protest in Peru*. Durham, NC: Duke University Press.

Steffen, Will, Paul J. Crutzen, and John R. McNeill. 2007. "The Anthropocene: Are Humans Now Overwhelmingly the Greatest Forces of Nature." *Ambio* 36: 61–62.

Stengers Isabelle. 2010. *Cosmopolitics I*. Minneapolis: University of Minnesota Press.

———. 2011. *Cosmopolitics II*. Minneapolis: University of Minnesota Press.

Stensrud, Astrid. 2010. "Los peregrinos urbanos en Qoyllurit'i y el juego mimético de miniaturas." *Anthropologica* 28: 39–65.

———. 2016a. "Climate Change, Water Practices and Relational Worlds in the Andes." *Ethnos* 81 (1): 75–98.

———. 2016b. "Harvesting Water for the Future Reciprocity and Environmental Justice in the Politics of Climate Change in Peru." *Latin American Perspectives* 43 (4): 56–72.

———. 2016c. "Dreams of Growth and Fear of Water Crisis: The Ambivalance of 'Progress' in the Majes-Siguas Irrigation Project, Peru." *History and Anthropology* 27 (5): 569–584.

———. 2017. "Precarious Entrepreneurship: Mobile Phones, Work and Kinship in Neoliberal Peru." *Social Anthropology* 25 (2): 159–173.

———. 2019. "Formalization of Water Use and Conditional Ownership in Colca Valley, Peru." *Water Alternatives* 12 (2): 521–537.

———. 2021. *Watershed Politics and Climate Change in Peru*. London: Pluto Press.

Stern, Steve. 1993. *Peru's Indian Peoples and the Challenge of Spanish Conquest, Huamanga to 1640*. Madison: University of Wisconsin Press.

Strang, Veronica. 2005. "Common Senses: Water, Sensory Experiences and the Generation of Meaning." *Journal of Material Culture* 10 (1): 92–120.

———. 2015. *Water: Nature and Culture*. London: Reaktion Books.

———. 2016. "Infrastructural Relations: Water, Political Power and the Rise of a New 'Despotic Regime.'" *Water Alternatives* 9 (2): 292–318.

Strauss, Sarah, and Ben Orlove, eds. 2003. *Weather, Climate and Culture*. Oxford: Berg.

Swyngedouw, Erik. 2006. "Circulations and Metabolisms: (Hybrid) Natures and (Cyborg) Cities." *Science as Culture* 15 (2): 105–121.

———. 2009. "The Political Economy and Political Ecology of the Hydrosocial Cycle." *Journal of Contemporary Water Research & Education* 142: 56–60.

Szeminski, Jan. 1984. *La utopía tupacamarista*. Lima: Pontificia Universidad Católica del Perú.

Taussig, Michael. 1980. *The Devil and Commodity Fetishism in South America*. Chapel Hill: University of North Carolina Pres.

Thurner, Mark. 1997. *From Two Republics to One Divided: Contradictions of Post-colonial Nationmaking in Andean Peru.* Durham, NC: Duke University Press.

Tortajada, Cecilia, ed. 2015. *Integrated Water Resources Management: From Concept to Implementation.* London: Routledge.

Tsing, Anna Lowenhaupt. 2015. *The Mushroom at the End of the World: On the Possibility of Life in Capitalist Ruins.* Princeton, NJ: Princeton University Press.

Trawick, Paul. 2003. *The Struggle for Water in Peru: Comedy and Tragedy in the Andean Commons.* Stanford, CA: Stanford University Press.

Treacy, John M. 1994. *Las chacras de Coporaque: Andenería y riego en el valle del Colca.* Lima: Instituto de Estudios Peruanos.

Triscitti, Fiorella. 2013. "Mining, Development and Corporate-Community Conflicts in Peru." *Community Development Journal* 48 (3): 437–450.

Ullberg, B. Susann. 2019. "Making the Megaproject: Water Infrastructure and Hydrocracy at the Public-Private Interface in Peru." *Water Alternatives* 12 (2): 503–520.

Urteaga, Patricia, and Rutgerd Boelens, eds. 2006. *Derechos colectivos y políticas hídricas en la región andina.* Lima: IEP (Instituto de Estudios Peruanos).

Urton, Garry. 1981. *At the Crossroads of the Earth and the Sky: An Andean Cosmology.* Austin: University of Texas Press.

USAID and Catie (Solutions for Environment and Development). 2016. *Diagnóstico de la subcuenca del río Shullcas.* Huancayo: USAID.

USGS (US Geological Survey). 2018a. *Ice, Snow, and Glaciers: The Water Cycle.* https://water.usgs.gov/edu/watercycleice.html.

———. 2018b. *Glaciers and Icecaps: Storehouses of Freshwater.* https://water.usgs.gov/edu/earthglacier.html.

Valderrama, Ricardo, and Carmen Escalante.1988. *Del Tayta Mallku a la Mama Pacha.* Lima: Desco (Centro de Estudios y Promoción del Desarrollo).

van den Berghe, Pierre, and George Primov. 1977. *Inequality in the Peruvian Andes: Class and Ethnicity in Cuzco.* Columbia: University of Missouri Press.

Vera Delgado, Juana, and Linden Vincent. 2013. "Community Irrigation Supplies and Regional Water Transfers in the Colca Valley, Peru." *Mountain Research and Development* 33 (3): 195–206.

Vergara, Walter, Alejandro Deeb, Adriana Valencia, Raymond Bradley, Bernard Francou, Alonso Zarzar, Alfred Grünwald, and Seraphine Haeussling. 2007. "Economic Impacts of Rapid Glacier Melt in the Andes." *EOS, Transactions, American Geophysical Union* 88 (25): 261–268.

Vos, Jeroen. 2005. "Understanding Water Delivery Performance in a Large-Scale Irrigation System in Peru." *Irrigation and Drainage* 54 (1): 67–78.

Vuille, Mathias, Mark Carey, Christian Huggel, Wouter Buytaert, Antione Rabatel, Dean Jacobsen, Alvaro Soruco, Marcos Villacis, Christian Yarleque, Oliver E. Timm, Thomas Condom, Nadine Salzmann, and Jean-Emmanuel Sicart. 2018. "Rapid Decline of Snow and Ice in the Tropical Andes—Impacts, Uncertainties and Challenges Ahead." *Earth-Science Reviews* 176: 195–213.

World Bank. 2018. *World Bank Open Data.* https://data.worldbank.org/country/peru.

Yao, Tandong, Lonnie Thompson, Wei Yang, Wusheng Wu, Yang Yao, Xuejun Guo, Xiaoxin Yang, Keqin Duan, Huabiao Zhao, Baiqing Xu, Jiancheng Pu, Yang Xiang, Dambary Kattel, and Daniel Joswiak. 2012. "Different Glacier Status with Atmospheric Circulations in Tibetan plateau and Surrounding." *Nature Climate Change* 2: 663–667.

Yearly, S. 2009. "Sociology and Climate Change after Kyoto: What Role of Social Science in Understanding Climate Change." *Current Sociology* 57 (3): 389–405.

Zuidema Tom. 1986. "Inka Dynasty and Irrigation: Another Look at Andean Concepts of History." In *Anthropological Histories of Andean Polities*, edited by John Murra, Nathan Watchel, and Jacques Revel, 177–200. Cambridge: Cambridge University Press.

INDEX

Peregrinos al Sanctuario de Quyllur
Rit'i - Sinak'ara, 125
Peru, 2; Andean climate ethnography,
crafting of, 14–19, 15*map*; climate vulnerability, 154n5; emigration from, 30;
gamonalismo and indios, 28–31; Incas,
Spaniards, and huakas (holy places),
28–31; water crisis in, 31–33
pichuhuira, 58
Pigs for the Ancestors (Rappaport), 11–12
pilgrimages, 13, 15, 16–17, 18, 21, 23, 31,
40–42, 144, 147. *See also* ritual
customs; specific mountain names
Plan de Salvaguardia del Patrimonio
Cultural de Qoyllur Rit'i, 137
Plan Maestro del Àrea de Conservación
Regional Huaytapallana, 113
political ontology, 7
politics: of the Anthropocene, 5–6; Cabanaconde, history of, 71–72; climate
ethnography, crafting of, 14–19, 15*map*;
cosmopolitics, 6–9; gamonalismo and
indios, 28–31; in Inca society, 26; mountains as icons of Peru's political system,
37–38; in Peru, 2; political violence in
late 1980s and early 1990s, 97–98; of
Spanish colonial rule, 26–28; Tapay,
history of, 45–48
pollution: on Huaytapallana, 99–100, 102,
113; Karl Marx on, 9–10; from mining,
101, 102. *See also* trash, religious rituals and
population movement. *See* migration
postnormal science, 4–5
power, social and political: gamonalismo
and indios, 28–31; water, role of, 25–28;
water-power conundrum, 145–47. *See
also* water infrastructure and
management
PROFUDUA, 52
progresar, 46, 48
puna, 48
Puno, 74, 127
Puquio, 55
puquios (springs), 49, 54–55
pututo, 106, 107

Qamawara, 128–29
q'apas, 83

qarakuna, 38
qochayuyo (seaweed), 58–59
qolqelibro, 58
qorilibro, 58
Qoyllur Rit'i fiesta, 124–25, 126
Quechua, 1, 36
Quispicanchi, 123, 125, 132, 141, 142,
167–68n5
Quyllurit'i. *See* Mount Qulque Punku
(Quyllurit'i)

Racracalla, 100
rain capture and storage, 102
rainy season, 80
Ramírez, Juan Andrés, 132–33
Rapaz, 37, 38
Rappaport, Roy, 9, 13, 151
reducciones, 26–27, 39, 71
regidor (water allocator), 43, 45, 50, 159n12
regidores, 49
Reinhard, Johan, 81–82
religion: Evangelists, 126–27; gamonalismo
and indios, 28–31; ice, power of, 24–25;
under Inca rule, 25–28; under Spanish
Colonial rule, 26–28; syncretism and,
123–24, 126, 143–44. *See also* cosmology; ritual customs
remittances, 73
ritual customs, 1–2; Andean mountains
and, 34–39; in Cabanaconde, 83–84;
Cabanaconde, changes in, 70–71,
92–93; Cabanaconde, fiesta of Virgen
del Carmen, 74; climate ethnography,
crafting of, 14–19, 15*map*; death during,
133–34, 141–42, 147; of Incas, 25–27,
123, 128; in La Campiña, 84; mountain
pilgrimages, 40–41; on Mount Huaytapallana, 95–100, 102–9, 103*fig*, 104*fig*,
105*fig*, 106*fig*, 107*fig*, 109*fig*, 111–13; on
Mount Huaytapallana, climate change
adaptation and, 114–19; Mount
Quyllurit'i, adaptation compromises,
140–44; Mount Quyllurit'i, climatic
threats and, 136–40; Mount Quyllurit'i,
metaphorical metabolism, 131–35; on
Mount Quyllurit'i, 120–31; Mummy
Juanita, Inca rituals, 81–83; Qoyllur
Rit'i fiesta, 124–25; as self-regulating

ritual customs (*continued*)
thermometer, 150–51; under Spanish colonial rule, 26–28; specialists in, 115–17; in Tapay, 43–44, 45, 53–61, 56*tab*, 57*fig*, 58*fig*, 63–66, 84; as transactional metabolism, 18–19; tribute payments, 38–39; water metabolism and, 9, 13, 34–39
Roncoli, Carla, 16

sacerdote andino (Andean priest), 115
sacrifice: of humans, 26–27, 81–82, 92, 134, 142, 146, 164n28; as sacred ritual, 34, 37
Salas Carreño, Guillermo, 126, 127, 131, 136, 141, 144
Sallnow, Michael, 123
Salomon, Frank, 37
San Clemento, 55
San Francisco, 55
San Juan, 54
Santa María Magdalena, 54
Santa Marta, 55
Santa Rosa de Lima, 55
seawater, 58–59
SEDAM (Empresa de Servicios de Agua Potable y Alcantarillado Municipal), 101–2
Señor de Milagros, 123
Señor de Quyllurit'i, 123, 124, 125
Señor de Tayankani, 124
Seprigina River, 50
Sequeiros, Pablo Concha, 21, 120–21, 131, 136
sewage, on Mount Quyllurit'i, 136, 137
Shining Path, 97
Shining Snow, 131
Shullcas River, 100–102, 110
silver, 58
Sinak'ara, 121*fig*, 122, 125, 127, 136, 138–39
sociopolitical contexts: climate ethnography, crafting of, 14–19, 15*map*
Spanish colonial rule: Cabanaconde, history of, 71–72, 72*fig*; ritual customs, Mount Quyllurit'i, 123–31; water as power in, 26–28
starfish, 58–59
Stengers, Isabelle, 6
strategically situated ethnography, 14

surface watering, 89
syncretism, religious, 123–24, 126, 143–44

Tahuantinsuyo, 125
Tampoña River, 50
Tapay, 14, 15*map*, 17–18, 39–40, 45*fig*; brinco water allocation method, 89; Cabanaconde, comparison to, 75, 80, 93; climate change, effects of, 59–63; climate change adaptations, 63–66, 148; cosmology and rituals of, 53–59, 56*tab*, 57*fig*, 58*fig*; demographics of, 157–59nn1–6; fieldwork in, 19–22; gold mining in, 62–63, 64, 65, 66–67; history of, 45–48; migration from, 44–45; modernization in, 62–63, 67–68; regulatory and technological changes in, 62–63; ritual customs in, 43–44, 45, 53–61, 56*tab*, 57*fig*, 58*fig*, 63–67, 84; taxes and tariffs in, 49–50, 52–53; tourism in, 61–62, 66; water infrastructure and management in, 48–53, 51*fig*, 146–47, 159n16
tariffs. *See* taxes and tariffs
Tawantisuyo, 132, 141
taxes and tariffs: in Cabanaconde, 80; evasion of payment, 52–53; as gift to realpolitical, 83–84; in Tapay, 49–50, 52; on water, 18, 31–33; water law of 2009, 40; water-power conundrum, 146–47
tayanka, 124
Tayankani, 124
taytacha (little father), 124, 130
Taytacha Quyllurit'i, 123
tikets, 80
t'inka, 38
Tobias, Robert, 149
Tocallo, 55, 56
Todos Santos (All Saints Day), 43–44, 55–56
Tomanta, 79
tomas (offtakes), 49
tourism, 66; in Colca Valley, 61–62; impact on Mount Huaytapallana, 94–96, 98–100, 103–9, 103*fig*, 104*fig*, 105*fig*, 106*fig*, 107*fig*, 109*fig*, 111–13; impact on Mount Quyllurit'i, 136–40; Mummy Juanita and, 81–82

Founded in 1893,
UNIVERSITY OF CALIFORNIA PRESS
publishes bold, progressive books and journals
on topics in the arts, humanities, social sciences,
and natural sciences—with a focus on social
justice issues—that inspire thought and action
among readers worldwide.

The UC PRESS FOUNDATION
raises funds to uphold the press's vital role
as an independent, nonprofit publisher, and
receives philanthropic support from a wide
range of individuals and institutions—and from
committed readers like you. To learn more, visit
ucpress.edu/supportus.